HANDBOOK
OF GENETIC
ALGORITHMS

HANDBOOK
OF GENETIC
ALGORITHMS

Edited by
Lawrence Davis

VNR VAN NOSTRAND REINHOLD
New York

Library of Congress Catalog Card Number 90-12823
ISBN 0-442-00173-8

Manufactured in the United States of America

Published by Van Nostrand Reinhold
115 Fifth Avenue
New York, New York 10003

Chapman and Hall
2-6 Boundary Row
London, SE 1 8HN

Thomas Nelson Australia
102 Dodds Street
South Melbourne 3205
Victoria, Australia

Nelson Canada
1120 Birchmount Road
Scarborough, Ontario M1K 5G4, Canada

16 15 14 13 12 11 10 9 8 7 6 5 4 3

Library of Congress Cataloging-in-Publication Data

Davis, Lawrence.
    Handbook of genetic algorithms / Lawrence Davis.
        p.   cm.
    Includes bibliographical references and index.
    ISBN 0-442-00173-8
        1. Combinatorial optimization. 2. Algorithms. I. Title.
    QA402.5.D3   1991
    511'.6—dc20   90-12823
                                                   CIP

Preface

The field of genetic algorithms has been growing since the early 1970s, but only recently has it yielded real-world applications that demonstrate commercial potential. This point in a technology's life cycle—the point at which commercial applications of a technology begin to succeed—is an important one, for it is the point at which researchers are asked to explain the technology to others interested in applying it.

The field of genetic algorithms is certainly at that stage. Those of us who have worked in the field for some time are beginning to be paid to apply genetic algorithms to real problems—a sure sign that a field has gone beyond the theoretical stage. Workers in other fields are beginning to notice genetic algorithms and to observe that they are different from other algorithms in intriguing ways. Genetic algorithm sessions have been held at conferences on machine learning, artificial intelligence, neural networks, and operations research, with strong participation from people in those fields. Attendance at genetic algorithm conferences has grown impressively at each of the biennial international conferences. At the 1989 conference presentations were made by several persons who had applied the technology to their problems without interaction with people in the field—another sign that a technology has passed into the applications phase.

Much of the interest in genetic algorithms comes from people who wish to solve real-world problems, who are not genetic algorithm researchers, and who may be uncomfortable with the formal notation that is used in presenting research results. It is, then, a good time to produce a volume like this one, designed to explain to an interested reader, in nonmathematical terms, what genetic algorithms are about and how to apply them.

Because the genetic algorithm field is changing rapidly, I will attempt to communicate both the capabilities of the genetic algorithm as it currently stands and the areas that are the subject of current research. Some of the sections in this book will soon become outdated. In five years' time many of the papers in the second section—currently representative of the cutting edge of genetic algorithm application—may well be superseded by newer results. The historical part of the tutorial in Part 1 may survive five years' changes, but its later sections will certainly be outmoded by better techniques for carrying out the optimizations described there. With regard to the computer systems in Part III, suffice it to say that no computer system that is designed to be used is ever completed.

This book is somewhat restricted in scope. The genetic algorithm field has three major thrusts: research into the basic genetic algorithm, optimization using the genetic algorithm, and machine learning with classifier systems. The research thrust is well described in Goldberg (1989). The classifier system work is also described in Goldberg's text, as well as in Holland et al. (1986). This book does not deal with those two areas. Instead, it is directed to the optimization field, and its goal is to teach a reader unfamiliar with genetic algorithms how to apply them effectively to optimization problems. For more formal presentation of the field, and for descriptions of the interesting machine learning capabilities of classifier systems, you are referred to those two other texts.

This book owes much to the research and encouragement of many people whom I wish to thank here.

The most important acknowledgment is that of John Holland, who created the genetic algorithm field. The field would not exist if he had not decided to harness the power inherent in genetic processes in the early 1970s and functioned as the technical and political leader of the genetic algorithm field from its inception to the present time. Our understanding of the unique features of genetic algorithms has been shaped by the careful and insightful work of Holland and his students from the field's critical first years to the present time.

As interest in genetic algorithms has grown, other researchers—students of Holland, their students, and researchers from other backgrounds—have increased our store of knowledge about the performance of genetic algorithms. This book can only touch on the work that these researchers have done.

Our ability to apply genetic algorithms to real-world problems has improved greatly over the past ten years. This book would not be possible without the body of practical expertise produced by the many researchers who have adapted genetic algorithms to their own problems. Some of those applications are described in Part II of this book.

I have recently started my own consulting practice in the genetic algorithm field, but Bolt Beranek and Newman Inc. (BBN) was my employer while this book was being written. BBN has assisted me in many ways during the execution of this project. Frank Heart, Ray Nickerson, and Ed Walker took a personal interest in the project and provided resources that made the process of writing and editing smoother and more enjoyable.

This book uses a computer system—OOGA—to illustrate the concepts explained in the tutorial section. Dan Cerys of BBN has made OOGA a system that can be run in any object-oriented Common Lisp environment. If you use the system in your environment, you will appreciate the work Dan has done in this regard.

Marjan Bace, formerly of Van Nostrand Reinhold, first proposed this project to me. Marjan was a strong advocate of the idea of the book, which was a critical factor in my decision to forego research for a year in order to produce the book.

Steve Chapman of Van Nostrand Reinhold has managed the project on the publisher's end from its early stages to its completion. Steve's efforts have helped to smooth the production process.

Betty Pessagno, a Van Nostrand Reinhold editor, labored mightily on the style and uniformity of the book. Her efforts have greatly improved the book's clarity, although it has undergone some rewriting since she reviewed it and some infelicities may have crept back in.

Paul Milazzo of BBN designed the LaTeX format of this book in conjunction with Van Nostrand Reinhold and also provided a good deal of technical and aesthetic support. Paul and Marjan are currently writing a book explaining how to design and publish technical books using LaTeX. Paul's dissertation described a novel computer representation of books and its use in automated page layout. Many of the figures in this book were submitted electronically; Paul's achievements in integrating them into the book were heroic. Eight formatting tools were used to produce these figures; integrating the results of each required wizardry, patience, and long hours. Paul's efforts in this regard were much greater than we had foreseen, and we authors are grateful for the forbearance and attention Paul gave our work.

Sections of this book have been reviewed by members of the genetic algorithm field, including Lashon Booker, John Grefenstette, Gunar Liepins, Darrell Whitley, and Gilbert Syswerda. They did a very thorough job but I am responsible for the errors we didn't catch, especially those introduced after their review.

My own work in the genetic algorithm field has been strongly influenced by my interactions with my colleagues at Texas Instruments and Bolt Beranek and Newman Inc. Many of my implementations of genetic algorithms have involved collaborations with colleagues. At Texas Instruments, I worked primarily on semiconductor layout with Derek Smith. At BBN, I have worked on communication network design with Susan Coombs; on scheduling problems and simulated annealing with Frank Ritter; on training neural networks for passive sonar problems with David Montana; on hybrid neural network training methods with Wayne Mesard; on database discovery techniques with James Kelly; on scheduling with Gilbert Syswerda; and on statistical analysis of genetic algorithm performance with Herbert Gish. I have also learned a good deal from Tony Cox and Yuping Qiu of U S West throughout our collaboration on telephone network routing problems. Finally, I have enjoyed and benefited from discussions with Gilbert Syswerda of BBN and Stewart Wilson of the Rowland Institute for Science.

The writing of the tutorial in Part I was the most significant task for me in producing this book. The tutorial will always remind me of Chalk Creek Canyon in Colorado, where I spent two weeks in December of 1989 working on the first draft. During that time I received support in the form of plumbing advice, snow plowing, companionship on demand, and elk steaks from Hugh and Sally Pape, who are living out the dream of many of us acquainted with the Colorado mountain country. One could not ask for better neighbors during an intensive writing project.

Finally, there were Wendy and Alex, Jim and Merilyn, and Hal and Kirby—family members who have always supported and promoted the odd hours and long nights that genetic algorithm projects can require.

How to Use this Book

Study of this book will acquaint you with the theoretical and practical power of genetic algorithms. It will also provide you with the information you need to apply that power to problems of your own. The book has three parts, and you will probably wish to use each part differently.

Part I is a tutorial on the application of genetic algorithms. I have written it for the reader who has a general scientific interest in genetic algorithms or a problem to solve that the genetic algorithm may be suited for. My goal in the tutorial is to leave the reader with a feeling for how a genetic algorithm works, what problems genetic algorithms have been used for, and how to tailor them to a variety of other problems. The tutorial is best read sequentially. The tutorial references OOGA, the Object-Oriented Genetic Algorithm. OOGA is a computer system that embodies and exemplifies the concepts explained in the tutorial. The performance of OOGA is described in the text, but readers who wish to observe the system in action on a computer or to tailor it to their own problems will probably derive a fuller appreciation of the concepts being explained.

Part II contains 13 chapters by authors who have applied genetic algorithms to real problems. Part II will provide you with a variety of application ideas to draw on. Chapter 9 surveys the application types and genetic algorithm techniques used in each chapter in Part II. Part II may be read in any order. Using Chapter 9 as a guide, you may wish to find those chapters describing applications most like the problems you are interested in and begin your reading of the section there.

Part III consists of a computer diskette and documentation, available at additional cost by mail. The diskette contains OOGA and GENESIS, the genetic algorithm system written in C by John Grefenstette. GENESIS has been the standard distributed version of the genetic algorithm since Grefenstette began making it available to the public. After reading the first few chapters of the tutorial and the documentation provided on the diskette, you will be able to apply GENESIS and OOGA to a variety of problems, if you can write or modify software written in C or Common Lisp.

The computer diskette is optional. You do not need to run OOGA in order to understand the tutorial, although doing so may give you a better feeling for the material. None of the chapters in Part II is related to the computer code. The best way to use Part III is probably to work through the tutorial examples in OOGA as they arise in the text, and then to experiment with OOGA or GENESIS on problems of your own. Your choice of which system to use will probably be based on a number of factors, including the computer language you have available to you and the type of problems you intend to attack with a genetic algorithm. In the final chapter of the book these computer systems are discussed in more detail and an order form is enclosed.

Contents

Preface   v

I   A Genetic Algorithms Tutorial   xii

1   What is a Genetic Algorithm?   1

2   Performance Enhancements   23

3   Further Evolution of the Genetic Algorithm   43

4   Hybrid Genetic Algorithms   54

5   Hybridization and Numerical Representation   61

6   Order-Based Genetic Algorithms and the Graph Coloring Problem   72

7   Parameterizing a Genetic Algorithm   91

8   Where to Go From Here?   99

II Application Case Studies 103

9 Overview of Part II 104

10 Genetic Algorithms in Parametric Design of Aircraft 109

11 Dynamic Anticipatory Routing in Circuit-Switched Telecommunications Networks 124

12 A Genetic Algorithm Applied To Robot Trajectory Generation 144

13 Genetic Algorithms, Nonlinear Dynamical Systems, and Models of International Security 166

14 Strategy Acquisition with Genetic Algorithms 186

15 Genetic Synthesis of Neural Network Architecture 202

16 Air-Injected Hydrocyclone Optimization via Genetic Algorithm 222

17 A Genetic Algorithm Approach to Multiple-Fault Diagnosis 237

18 A Genetic Algorithm for Conformational Analysis of DNA 251

19 Automated Parameter Tuning for Interpretation of Synthetic Images 282

20 Interdigitation: A Hybrid Technique for Engineering Design Optimization Employing Genetic Algorithms, Expert Systems, and Numerical Optimization 312

21 Schedule Optimization Using Genetic Algorithms 332

22 The Traveling Salesman and Sequence Scheduling: Quality Solutions Using Genetic Edge Recombination 350

III GENESIS and OOGA: Genetic Algorithm Software 373

23 Concerning GENESIS and OOGA 374
Contributing Authors 378
Index 381

Part I

A Genetic
Algorithms Tutorial

1

What is a Genetic Algorithm?

INTRODUCTION

In this tutorial I am going to tell you as much as I can, subject to space constraints, about genetic algorithms and their application to real-world problems. We will barely begin to consider many important subjects and we will ignore many others entirely, but when you have finished the tutorial I believe you will be able to apply genetic algorithms to problems of your own. You will also be prepared to continue reading other books and papers in the field.

The tone of this tutorial is deliberately personal. I wrote the first draft during a wintry two-week period in a Colorado mountain cabin. While working out the draft, I imagined that I was talking to a colleague about genetic algorithms, with a computer at hand that demonstrated the points we were discussing. It may be useful for you to read the tutorial pretending that the same conditions obtain.

This first chapter of the tutorial discusses several major points. First, we consider some features of biological evolution that inspired John Holland's invention of genetic algorithms. Next, we consider general features of genetic algorithms. Next we move to a more specific implementation of these ideas—the architecture of OOGA, a computer implementation of the genetic algorithm idea. Finally, we examine the inner workings of the OOGA genetic algorithm. When the chapter is completed, we will have considered four variations on the genetic algorithm theme, each at a different remove from the phenomena of biological evolution.

NATURAL EVOLUTION: THE INITIAL INSPIRATION

Genetic algorithms were invented to mimic some of the processes observed in natural evolution. In this section I note what those processes were and discuss why John Holland, the field's inventor, thought it would be a good idea to incorporate them in an algorithm.

Biologists have been intrigued with the mechanics of evolution since the evolutionary theory of biological change gained acceptance. Many people, biologists included, are astonished that life at the level of complexity that we observe could have evolved in the relatively short time suggested by the fossil record.

The mechanisms that drove this evolution are not fully understood, but some of its features are known. Evolution takes place on *chromosomes*— organic devices for encoding the structure of living beings. A living being is created partly through a process of *decoding* chromosomes. The specifics of chromosomal encoding and decoding processes are not fully understood, but here are some general features of the theory that are widely accepted:

- Evolution is a process that operates on chromosomes rather than on the living beings they encode.

- Natural selection is the link between chromosomes and the performance of their decoded structures. Processes of natural selection cause those chromosomes that encode successful structures to reproduce more often than those that do not.

- The process of reproduction is the point at which evolution takes place. Mutations may cause the chromosomes of biological children to be different from those of their biological parents, and recombination processes may create quite different chromosomes in the children by combining material from the chromosomes of two parents.

- Biological evolution has no memory. Whatever it knows about producing individuals that will function well in their environment is contained in the gene pool—the set of chromosomes carried by the current individuals—and in the structure of the chromosome decoders.

These features of natural evolution intrigued John Holland in the early 1970's. Holland believed that, appropriately incorporated in a computer algorithm, they might yield a technique for solving difficult problems in the way that nature has done—through evolution. And so he began work on algorithms that manipulated strings of binary digits—1's and 0's—that he called *chromosomes*. Holland's algorithms carried out simulated evolution on populations of such chromosomes. Like nature, his algorithms solved the problem of finding good chromosomes by manipulating the material in the chromosomes blindly. Like nature, they knew nothing about the type of problem they were solving. The only information they were given was an evaluation of each chromosome they produced, and their only use of that evaluation was to bias the selection of chromosomes so that those with the best evaluations tended to reproduce more often than those with bad evaluations.

These algorithms, using simple encodings and reproduction mechanisms, displayed complicated behavior, and they turned out to solve some extremely difficult problems. Like nature, they did so without knowledge of the decoded world. They were simple manipulators of simple chromosomes. Yet when we use the descendants of those algorithms today, we find that they can evolve better designs, find better schedules, and produce better solutions to a variety of other important problems that we cannot solve as well using other techniques.

When Holland first began to study these algorithms, they did not have a name. As the field began to demonstrate its potential, however, it was necessary to christen it. In reference to its origins in the study of genetics, Holland named the field *genetic algorithms*.

You should know that, although the findings of evolutionary biologists inspired the field of genetic algorithms in its early years, and although the findings of biologists and geneticists continue to influence the field somewhat, this influence is for the most part unidirectional. I know of no genetic algorithm application in the area of genetics, nor, to my knowledge, have the findings in our field impacted the theories of biologists. In this regard, genetic algorithms seem to be like neural networks and simulated annealing—other algorithms based on powerful metaphors from the natural world. Natural phenomena inspired these algorithmic abstractions, but scientists who study the natural phenomena have not as yet been greatly influenced by the algorithmic abstractions. Instead, the abstract fields have become disciplines in their own right after the initial moments of metaphorical inspiration, and the genetic algorithm, neural network, and simulated annealing fields at least have subsequently wandered far from the disciplines that inspired them. In the case of genetic algorithms, you will see later in this book just how far they have wandered.

A TOP-LEVEL VIEW OF THE GENETIC ALGORITHM

We can abstract natural phenomena into an algorithm in many ways. In this section we will see how Holland embodied the preceding features of the theory of natural evolution in his genetic algorithms.

To begin with, let us consider the mechanisms that link a genetic algorithm to the problem it is solving. There are two such mechanisms—a way of *encoding* solutions to the problem on chromosomes, and an *evaluation function* that returns a measurement of the worth of any chromosome in the context of the problem.

Let us discuss these mechanisms in more detail.

The technique for encoding solutions may vary from problem to problem and from genetic algorithm to genetic algorithm. In Holland's work, and in the work of most of his students, encoding is carried out using bit strings. We will use bit strings in the genetic algorithm described in this chapter, but genetic algorithm researchers have used many other types of encoding techniques. Probably no one technique works best for all problems, and a certain amount of art is involved in selecting a good decoding technique when a problem is being attacked. One important question which we take up in this tutorial centers on what factors we should consider when selecting a representation technique in the context of a real-world problem.

The evaluation function is the link between the genetic algorithm and the problem to be solved. An evaluation function takes a chromosome as input and returns a number or list of numbers that is a measure of the chromosome's performance on the problem to be solved. Evaluation functions play the same role in genetic algorithms that the environment plays in natural evolution. The interaction of an individual with its environment provides a measure of its fitness, and the interaction of a chromosome with an evaluation function provides a measure of fitness that the genetic algorithm uses when carrying out reproduction.

Given these initial components—a problem, a way of encoding solutions to it, and a function that returns a measure of how good any encoding

> **The Genetic Algorithm**
> 1. Initialize a population of chromosomes.
> 2. Evaluate each chromosome in the population.
> 3. Create new chromosomes by mating current chromosomes; apply mutation and recombination as the parent chromosomes mate.
> 4. Delete members of the population to make room for the new chromosomes.
> 5. Evaluate the new chromosomes and insert them into the population.
> 6. If time is up, stop and return the best chromosome; if not, go to 3.

Figure 1.1: Top-level description of a genetic algorithm

is—we can use a genetic algorithm to carry out simulated evolution on a population of solutions. Figure 1.1 contains a top-level description of the genetic algorithm itself—the algorithm that uses these components to simulate evolution.

If all goes well throughout this process of simulated evolution, an initial population of unexceptional chromosomes will improve as parents are replaced by better and better children. The best individual in the final population produced can be a highly-evolved solution to the problem. In the rest of this tutorial we will consider a number of issues having to do with making this process happen efficiently and effectively.

THE ANATOMY OF A GENETIC ALGORITHM

The algorithmic description of a genetic algorithm in Figure 1.1 describes most genetic algorithms. But different researchers have implemented this description in different ways. In this section we move from the more abstract description of genetic algorithms to a particular one. In making this move, we leave the realm of universal applicability. In fact, the anatomy of a genetic algorithm given in this section and in the rest of the tutorial is my own. I don't know of any other researcher who breaks things down in just the same way. In working through this tutorial, then, you should remember that from this point on you are seeing one exemplification of the idea of a genetic algorithm. This exemplification is contained in the Object-Oriented Genetic Algorithm—OOGA—that is optionally available on computer diskette and that accompanies this tutorial. Another exemplification—the one that has been the standard in the field for the last five or six years—is John Grefenstette's GENESIS, also contained on the diskette. But however the genetic algorithm is described, the features discussed here will be included.

There are many components of a genetic algorithm. In OOGA these components are collected into three modules: an Evaluation Module, a Population Module, and a Reproduction Module. The following computer printout is a graphic representation of the architecture of GA 1-1, a genetic algorithm containing these modules. (Note that OOGA has labeled its modules and the techniques they contain in upper case letters. Parameter values are labeled in lower case.) This printout was obtained by typing (display *ga*) to a Lisp Listener after GA 1-1 had been created.

```
----------------------------------------------------------------

   ANATOMY OF GA-1-1

----------------------------------------------------------------

   EVALUATION-MODULE

      EVALUATION FUNCTION: BINARY-F6

----------------------------------------------------------------

   POPULATION-MODULE

      REPRESENTATION TECHNIQUE:  BINARY-REPRESENTATION
            Bit string length: 44
      INITIALIZATION TECHNIQUE:   RANDOM-BINARY-INITIALIZATION
      DELETION TECHNIQUE:  DELETE-ALL
      REPRODUCTION TECHNIQUE:  GENERATIONAL-REPLACEMENT
```

```
PARENT SELECTION TECHNIQUE:
        ROULETTE-WHEEL-PARENT-SELECTION
FITNESS TECHNIQUE:  FITNESS-IS-EVALUATION

PARAMETERIZATION TECHNIQUES: none

Population Size: 100
Desired Trials: 4000
Current Index: 4000

---------------------------------------------------------------


REPRODUCTION-MODULE

    OPERATOR SELECTION TECHNIQUE:  USE-FIRST-OPERATOR
    OPERATORS:
        ONE-POINT-CROSSOVER-AND-MUTATE
        Bit Mutation Rate: 0.008
        Crossover Rate: 0.65

    PARAMETERIZATION TECHNIQUES: none

    Operator Weights: (100)


-------------------------------------------
```

In this display you can see that GA 1-1 is composed of three principal parts—an Evaluation Module, a Population Module, and a Reproduction Module. The Evaluation Module contains an evaluation function that measures the worth of any chromosome on the problem to be solved. The Population Module contains a population of chromosomes and techniques for creating and manipulating that population. The Reproduction Module contains techniques for creating new chromosomes during reproduction.

GA 1-1 is the most elementary genetic algorithm we will consider in this tutorial. It will serve as a sort of proto-algorithm, in that all the genetic algorithms to come will be derived from successive replacements to and extensions of the parts of GA 1-1. In fact, nearly every technique in GA 1-1 will have been modified by the time the tutorial is over.

Let us examine the workings of the three modules of GA 1-1 in greater detail.

The Evaluation Module

The Evaluation Module of GA 1-1 contains an evaluation function. In GA 1-1, the evaluation function is binary f6, a mathematical function. You will frequently encounter such functions in the genetic algorithm literature. The operation of this function is not simple. Let us consider it here in some detail.

Binary f6 does three things. It decodes a chromosome that is a list of 44 bits by converting it into two real numbers; it plugs those real numbers into a mathematical function; and it returns the value of the function for those two numbers. This value that is returned will be the evaluation of the chromosome.

During the first, decoding stage binary f6 takes a chromosome that is a list of 44 bits and converts it into two real numbers, each of which lies in the range between -100 and 100. It does so in three steps. First, it interprets the initial 22 bits of the chromosome as a representation of an integer x_1 in base-2 notation and it interprets the last 22 bits as an integer y_1 in base-2 notation. It then multiplies x_1 and y_1 by .00004768372718899898, creating x_2 and y_2. This division step maps the range of possible values of x_1 and y_1 (a range from 0 to $2^{22} - 1$) to the range between 0 and 200. Finally, the evaluator subtracts 100 from x_2 and y_2 to create x_3 and y_3, real numbers between -100 and 100. Figure 1.2 illustrates this process. In the figure, a chromosome has been partitioned and interpreted, then divided by the constant and shifted to lie in the range from -100 to 100. The result is that a bit string chromosome—a list of 0's and 1's—is decoded to produce a pair of integers.

Now binary f6 computes a mathematical function using x_3 and y_3 as inputs. This function, an inverted version of the function called F6 in Schaffer, Caruana, Eshelman, and Das (1989), is the following:

$$0.5 - \frac{(\sin \sqrt{x^2 + y^2})^2 - 0.5}{(1.0 + 0.001(x^2 + y^2))^2}$$

Our goal is to find values of x_3 and y_3 which, substituted for x and y in this formula, produce the greatest possible value. More technically, our goal is to *optimize* this function.

Schaffer et al. chose binary f6 to measure the effectiveness of genetic algorithms because it has several features that make it an interesting test case. One is that the function has a single optimal solution, seen at the center of Figure 1.3. In making this graph, the y value of the function has been held fixed at its optimal point. The x value is varied from -100 to 100, and the height of the curve for any x value is the evaluation returned by f6.

Example of Decoding Routine Used By Binary F6

Chromosome:

000010100001100000000110001010100011101111011

is partitioned into:

000010100001100000001 and 1000101010001110111011

These bit strings are converted from base 2 to base 10 to yield x_1 and y_1:

165377 and 2270139

These numbers are multiplied by:

.00004768372718899898

to yield x_2 and y_2:

7.885791751335085 and 108.24868875710696

These numbers are diminished by 100 to yield x_3 and y_3:

-92.11420824866492 and 8.248688757106959

The evaluation of the chromosome is 0.5050708, the value of f6 when applied to x_3 and y_3.

Figure 1.2: Example of the decoding process used by binary f6

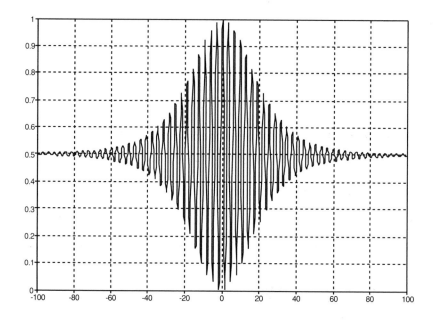

Figure 1.3: Graph of f6 varying x when y is held constant at its optimal value

given that x value and the optimal y value. You can see that the optimal value of the function occurs when the value of x is at 0. The function is symmetric in x and y, so the graph over y-values holding the x-value at its optimum would look the same. F6 is a function that describes a very hilly two-dimensional landscape, with an optimal region that occupies a tiny fraction of the total area.

Binary f6 is difficult for many function optimization strategies because of its oscillation. *Hill-climbing* techniques, that begin at a randomly generated point and proceed in the steepest direction until they can climb no more, will rapidly become trapped on a suboptimal oscillation. Only one of the hills has the optimal value at its top, and the chances of landing on that hill at random are not great.

There is nothing special about f6 that ties it to the rest of GA 1-1. We could substitute any other function as long as the new function took two real numbers as input. If the genetic algorithm was working well, it would evolve chromosomes that encoded optimal or near-optimal values of that function instead of f6.

This point is worth stressing. Nowhere except in the evaluation function is there any information in GA 1-1 about the problem to be solved. As far as GA 1-1 is concerned, it is merely reproducing and operating on bit strings so that those with higher evaluations tend to reproduce more often. The decoding procedure and the evaluation function could be replaced, and these changes would require no change to the other modules of the genetic algorithm, as long as the number of bits in the chromosomes stayed constant.

The Population Module

As you can see, the Population Module of GA 1-1 contains a number of techniques. In this section we consider the simpler ones.

The representation technique is binary representation. Each chromosome will be composed of a list of binary digits. We see that each list will contain 44 bits.

The initialization technique used is random binary initialization. The algorithm will create its starting population by filling it with randomly generated bit strings.

The deletion technique is delete all. All members of the old population will be deleted when reproduction has occurred.

The reproduction technique used is generational replacement. Each replacement of the old population by the new one is called a *generation* when this technique is used.

The parent selection technique used is roulette wheel parent selection. This technique is somewhat complicated, and is described in the next section.

The fitness technique is fitness is evaluation. GA 1-1 will use the evaluation of each chromosome as the measure of its reproductive fitness, without any modification.

There are no parameterization techniques. Techniques of this type are described in a later chapter.

With regard to parameter values, you can see that the population size is 100. That is, the population will always contain 100 individuals. The number of desired trials is 4000. GA 1-1 will create 40 generations of 100 individuals before stopping. The current index is 4000. This index is always set to equal the number of individuals that have been placed in the population at the point when the display is done. In this case, the genetic algorithm has displayed its state after a run has been completed.

The Reproduction Module

The Reproduction Module of GA 1-1 includes three component techniques and a parameter value.

The operator selection technique is use first operator. Since there is only one operator, this technique guarantees that it is the one to be used. More interesting techniques are available when more operators are being used.

The operators list has only one operator, one-point crossover and mutate. This operator has two parameter settings, a bit mutation rate of .008, and a crossover rate of .65. This operator and its parameter settings are described in more detail below. (You should note that this presentation of operators differs slightly from that in Goldberg (1989). There Goldberg refers to reproduction as an operator. Here only specific techniques for creating children that are different from their parents are called operators.)

There are no parameterization techniques.

The operator weights parameter is set at (100). This parameter will not be used until the next chapter.

GA 1-1 IN ACTION

When GA 1-1 is run, the Evaluation, Population, and Reproduction modules work together to effect the evolution of a population of chromosomes. First, the initialization technique generates a population of 100 random bit strings. This population makes up the *first generation* of chromosomes. Next, each chromosome in the population is evaluated by the evaluation function, and fitness values for each chromosome are computed by the fitness technique.

Now GA 1-1 begins a series of cycles of replacing its current population of chromosomes by a new population. Each of these cycles produces a new generation of chromosomes. In each cycle there are 50 occurrences of reproduction. In reproduction, we use the parent selection technique to pick two *parent* chromosomes. The Reproduction Module applies the one-point crossover and mutate operator to those two parents to generate two new chromosomes, called *children*. The operator may cause those new chromosomes to differ from their parents. After 50 occurrences of reproduction, 100 new chromosomes have been produced. These 100 new chromosomes are evaluated, their fitnesses are calculated, and they replace the current 100 chromosomes to form the next generation.

These generational cycles continue until 4000 individuals have been produced, at the fortieth generation. At this point, GA 1-1 halts. If all has gone well, a randomly generated population of chromosomes has evolved so that later generations contain individuals that are better than any found earlier. In the next chapter you will see a printout of a run of GA 1-1 showing that this is indeed what happens.

The interactions between the Population Module of GA 1-1 and the other two modules during a run are very simple. When GA 1-1 is running, the Population Module asks the Reproduction Module for a new population. The Reproduction Module asks the Population Module for parents to be used in reproduction events. In each reproduction event, the Reproduction Module gets two parents from the Population Module, applies the one-point crossover and mutate operator to the parents, and sends the two children created to the Population Module until enough children have been generated to fill a new generation of 100 chromosomes.

The Population Module interacts with the Evaluation Module during the run, in that whenever a new set of children has been produced, the Population Module asks the Evaluation Module to evaluate each new child before the child is placed in the population.

PARENT SELECTION, MUTATION, AND CROSSOVER

Several components of GA 1-1 require further discussion. In this section we examine them in more detail.

Roulette Wheel Parent Selection

The purpose of parent selection in a genetic algorithm is to give more reproductive chances, on the whole, to those population members that are the most fit. There are many ways to do this, and you will learn about several of them later. One commonly-used technique is roulette wheel parent selection, the technique used by GA 1-1. Roulette wheel parent selection is described in Figure 1.4.

Figure 1.5 shows a population of ten chromosomes with a set of evaluations totaling 76. The first row of the figure contains the index of each chromosome, the second contains each chromosome's fitness, and the third contains the running total of fitness. The figure also shows seven numbers generated randomly from the interval between 0 and 76, together with the index of the chromosome that would be chosen by roulette wheel parent selection for each of these numbers. In each case, the chromosome chosen

---

Roulette Wheel Parent Selection

1. Sum the fitnesses of all the population members; call the result total fitness.

2. Generate n, a random number between 0 and total fitness.

3. Return the first population member whose fitness, added to the fitnesses of the preceding population members, is greater than or equal to n.

Figure 1.4: The roulette wheel parent selection algorithm

is the first one at which the running total of fitness is greater than or equal to the random number.

The effect of roulette wheel parent selection is to return a randomly selected parent. Although this selection procedure is random, each parent's chance of being selected is directly proportional to its fitness. On balance, over a number of generations this algorithm will drive out the least fit members and contribute to the spread of the genetic material in the fittest population members. Of course, it is possible that the worst population member could be selected by this algorithm each time it is used. Such an occurrence would inhibit the performance of a genetic algorithm using this selection technique, but the odds of this happening in a population of any size are negligible.

This algorithm is referred to as roulette wheel selection because it can be viewed as allocating pie-shaped slices on a roulette wheel to population members, with each slice proportional to the member's fitness. Selection of a population member to be a parent can then be viewed as a spin of the wheel, with the winning population member being the one in whose slice the roulette spinner ends up.

However it is viewed, this parent selection technique has the advantage that it directly promotes reproduction of the fittest population members by biasing each member's chances of selection in accord with its evaluation. Note that when using this technique the fitness values of the chromosomes should be positive numbers.

Examples of Roulette Wheel Parent Selection

Chromosome	1	2	3	4	5	6	7	8	9	10
Fitness	8	2	17	7	2	12	11	7	3	7
Running Total	8	10	27	34	36	48	59	66	69	76

Random Number	23	49	76	13	1	27	57
Chromosome Chosen	3	7	10	3	1	3	7

Figure 1.5: Examples of roulette wheel parent selection. The first table shows the fitness of ten chromosomes and the running total of fitness. The second table shows the chromosome that would be chosen by the roulette wheel algorithm using these fitness values for each of seven randomly generated numbers.

One-Point Crossover And Mutate

One-point crossover and mutate is another component of GA 1-1 that merits discussion. Its function is to cause chromosomes created during reproduction to differ from those of their parents. First the operator makes a copy of each parent—the two children. Then it applies two functions to those children that may alter them. We consider those two processes in reverse order below.

Bit Mutation. Bit mutation is one procedure carried out by one-point crossover and mutate. When bit mutation is applied to a bit string it sweeps down the list of bits, replacing each by a randomly selected bit if a probability test is passed. Bit mutation has an associated probability parameter that is typically quite low. This parameter is set at .008 for GA 1-1. Thus, when bit mutation is applied to a chromosome during a run of GA 1-1, each bit in the chromosome will have an eight in one thousand chance of being randomly replaced.

Figure 1.6 contains several examples of the operation of bit mutation. There we see three parent chromosomes of length 5, the numbers randomly generated by the bit mutation probability check, the randomly generated bit that replaces the bit that passed the probability check, and the three chromosomes that result from the action of bit mutation. We see that for the first chromosome (as will be the case for most chromosomes of this length) the probability test is never passed, and so in this case the output of bit mutation is the same as the input. In the second case, the probability

Examples of Bit Mutation

Old Chromosome				Random Numbers				New Bit	New Chromosome			
1	0	1	0	.801	.102	.266	.373	–	1	0	1	0
1	1	0	0	.120	.096	**.005**	.840	0	1	1	0	0
0	0	1	0	.760	.473	.894	**.001**	1	0	0	1	1

Figure 1.6: Examples of bit mutation. The table shows three chromosomes of length four, random numbers generated for each bit in each chromosome, the new bits for the two occasions when the random number test is passed, and the resulting chromosomes. The two random numbers that cause a new bit to be generated are printed in bold face.

test is passed for the fourth bit. However, the randomly generated bit is the same as the original bit, and so there is no effective change. For the third chromosome, the probability test is passed for the fifth bit, and the new bit differs from the old. Thus, the third chromosome in the figure is the only one that is changed by the bit mutation operator.

There is a potential source of confusion having to do with the bit mutation rate. Some genetic algorithm practitioners use bit mutation to flip bits. Using this variant, if the probability test is passed, we replace a 1 by a 0, and a 0 by a 1. This approach results in an effective rate of mutation that is twice as high. (Recall that when mutation is applied to a bit using the bit mutation operator described here, half the time the new value will be the same as the old value, as happened in the second example in Figure 1.6.) You should be alert to this distinction if you attempt to replicate published results using this operator.

One-Point Crossover.   The idea of a process that produces mutations of chromosomes is probably familiar to you. Most of us learned about biological evolution in our early school years. In my biology classes I was taught that evolution is a process based on natural selection and random mutation. But this view of evolution is incomplete. There are other processes that alter chromosomes during reproduction, and some evolutionary biologists believe they may be at least as important as mutation. One of them is called *crossover*. In nature, crossover occurs when two parents exchange parts of their corresponding chromosomes. In a genetic algorithm, crossover recombines the genetic material in two parent chromosomes to make two children. John Holland experimented with a crossover operator

Examples of One-Point Crossover

Parent 1:   1   1   1   1 | 1   1   Child 1:   1   1   1   1   0   0
                                    $\Longrightarrow$
Parent 2:   0   0   0   0 | 0   0   Child 2:   0   0   0   0   1   1

Parent 1:   1   0   1 | 1   0   1   Child 1:   1   0   1   1   0   0
                                    $\Longrightarrow$
Parent 2:   0   0   1 | 1   0   0   Child 2:   0   0   1   1   0   1

Figure 1.7: Two examples of one-point crossover. The children are made by cutting the parents at the point denoted by the vertical line and exchanging parental genetic material after the cut.

called *one-point crossover*. One-point crossover occurs when parts of two parent chromosomes are swapped after a randomly selected point, creating two children. Figure 1.7 shows two examples of the application of one-point crossover during a run of a genetic algorithm.

One important feature of one-point crossover you should be aware of is that it can produce children that are radically different from their parents. The first example in Figure 1.7 is an instance of this. Another important feature is that one-point crossover will not introduce differences for a bit in a position where both parents have the same value. Thus, in the second example, we see that bit positions 2, 3, 4, and 5 have the same value in both parents and both children, even though crossover has occurred. An extreme instance occurs when both parents are identical. In such cases, crossover can introduce no diversity in the children.

Crossover is an extremely important component of a genetic algorithm. Many genetic algorithm practitioners believe that if we delete the crossover operators from a genetic algorithm the result is no longer a genetic algorithm. This claim has not been made for mutation operators. In fact, many genetic algorithm practitioners believe that the use of a crossover operator distinguishes genetic algorithms from all other optimization algorithms. Certain optimization algorithms—dynamic programming, for instance—maintain populations of individuals and apply mutation-like operations to them, preserving the best ones. There are other optimization algorithms inspired by the processes of biological evolution that maintain populations of solutions, use reproduction by fitness, and apply mutation operations to the individuals in the populations. Genetic algorithm researchers believe that the performance of those algorithms will not be as good, in general, as that of genetic algorithms, because genetic algorithms employ crossover and these other algorithms do not. This belief needs to be supported by empirical test. Nonetheless, genetic algorithm researchers have shown that when crossover is deleted from a genetic algorithm, performance is degraded on a variety of problems. What is also required is to show that when crossover is added to the arsenal of similar optimization algorithms their performance is improved.

Crossover occupies a special place in the hearts of genetic algorithm practitioners. It is regarded as the distinguishing feature of algorithms in our field and as a critical accelerator of the search process when a genetic algorithm runs. Why do crossover operators inspire such strong reactions? There are many ways to answer this question. Let us consider the one having to do with sex, another natural process inspiring strong reactions.

Sexual reproduction is the reproductive technique employed by the more complicated creatures on our planet. Yet, compared with processes of reproduction involving only one parent, sexual reproduction is an expensive way to make offspring. A species using sexual reproduction must have more than one type of individual. Those individuals must differ in important respects, and they must spend time and energy finding each other when reproduction is to occur. None of this overhead obtains for single-parent reproduction. Since sexual reproduction has won out in the arena of natural selection, there must be some respects in which this overhead is repaid.

One widely held view on this question is that sexual reproduction allows rapid combination of beneficial new traits in a way that cannot be duplicated by mutation. For example, consider two species of organisms occupying roughly the same ecological niche and having the same chromosomal structure, except that one—the Mutes—reproduces by splitting with mutation, and the other—the Crosses—uses sexual reproduction with crossover as well as mutation. Suppose that the mutation rate is the same for both species, and suppose further that there are two independent and quite beneficial mutations that will immensely improve these organisms' fitness in their environmental niche. Let's call these mutations mutation A and mutation B. Suppose further that each mutation has a one in a billion chance of occurring.

What happens when mutation A and mutation B occur in our two populations? For the Mutes, members with mutation A or mutation B will likely come to dominate the population, since these mutations will confer added fitness on their possessors. But it will be a long time before any individual comes to have both mutations, since an individual with mutation A can acquire mutation B only through mutation, and these mutations are quite

---

Schemata in a Bit String

Chromosome (01101) contains 32 schemata including:
                0##0#   #1###   011#1
                #####   01101   0#1#1

Figure 1.8: Examples of schemata in a single bit string

unlikely. For the Crosses it is a different matter. As soon as individuals with mutation A and mutation B come to dominate in the population, it becomes quite likely (with respect to a billion to one chance) that crossover will act to combine them. Just how likely this is depends on how the mutations are encoded, but it is clear that the odds greatly favor an individual in the Cross species possessing both mutation A and B much sooner than an individual in the Mute species. Assuming that the possessor of both these mutations and its descendants will come to dominate the niche, the result is likely to be that the Crosses will eliminate the Mutes.

This fable is a simplistic portrayal of the benefits of crossover. More formal and detailed treatments are widely found in the genetic literature. The point they make is that crossover acts to combine *building blocks* of good solutions from diverse chromosomes. Figure 1.8 shows a single chromosome of length 5 and several of the building blocks contained in that chromosome. Each building block is represented as a list made up of three characters—1, 0, and "#". A 1 or 0 at any position in the building block means that the chromosome must have the same value at that position for it to contain the building block. A "#" at any position in the building block means that the value of the chromosome at that position is irrelevant to determining whether the chromosome contains the building block. Thus, in order to contain a given building block, a chromosome must match the 1's and 0's on the building block exactly. The "#"s function as "don't cares."

Holland calls each building block a *schema*. His investigation into the success of genetic algorithms led him to conclude that genetic algorithms manipulate *schemata* when they run. In fact, Holland's Schema Theorem asserts precisely this. If we use a reproduction technique that makes reproduction chances proportional to chromosome fitness, then we can predict the relative increase or decrease of a schema s in the next generation of a generational genetic algorithm as follows. Let r be the average fitness of all chromosomes in the population containing s. Let n be the number of chromosomes in the population containing s. Let a be the average fitness of all chromosomes in the population. Then the expected number of occurrences of s in the next generation of the population is $n * r/a$, minus disruptions caused by mutation and crossover.

In effect, the Schema Theorem says that a schema occurring in chromosomes with above-average evaluations will tend to occur more frequently in the next generation, and one occurring in chromosomes with below-average evaluations will tend to occur less frequently (ignoring the effects of mutation and crossover). The details of the Schema Theorem and other theoretical results are treated in detail in Goldberg's textbook and in survey papers in the field.

This feature of genetic algorithms has been described by Holland as one of *intrinsic parallelism*, in that the algorithm is manipulating a large number of schema in parallel. The reproduction mechanisms together with crossover cause the best schemata to proliferate in the population, combining and recombining to produce high-quality combinations of schemata on single chromosomes.

You can expect a genetic algorithm with crossover to win out over a similar algorithm without crossover as long as these conditions obtain: populations contain diverse members; representative samples of different building blocks of good solutions are available in the population; the evaluation function reflects the contributions of these building blocks; and crossover is capable of putting them together.

We have seen that crossover operators are important components of genetic algorithms. In fact, many genetic algorithm practitioners believe that mutation is merely a device for reintroducing diversity into the population when a bit has accidentally taken on a single value everywhere, and they believe that crossover is the genetic workhorse, a high-performance search technique that acts rapidly to combine what is good in the initial population and that continues to spread good schemata throughout the population as a genetic algorithm runs.

This concludes our discussion of the one-point crossover and mutate operator. The operator takes two parents as input, copies them, and applies the mutate and one-point crossover processes to the parents if the relevant probability tests are passed.

THE INVERSION OPERATOR

In Holland (1975) there are three techniques for causing children to be different from their parents: crossover, mutation, and inversion. Inversion operates on a single chromosome. It inverts the order of the elements between two randomly-chosen points on the chromosome. While this operator was inspired by a biological process, it requires additional overhead and it has not in general been found to be useful in genetic algorithm practice. Thus, it is rarely employed at present. Some genetic algorithm practitioners believe that if we ever use chromosomes that are one or more orders of magnitude longer than the ones we employ today, inversion will prove to be useful. We will not use inversion in the genetic algorithms here.

SUMMARY

The principal points addressed in this chapter were the following:

- Genetic algorithms were invented by John Holland in the 1970s to mimic some features of natural evolution.
- Evaluation functions link genetic algorithms with the problems we are asking them to solve.
- Genetic algorithms use mutation and crossover to create children that differ from their parents.
- Genetic algorithms use parent selection techniques that mimic the process of natural selection, in that the fittest individuals tend to reproduce most often.
- Genetic algorithms manipulate *schemata*—building blocks of good solutions—that are combined through crossover and that spread in the population in proportion to the relative fitness of the chromosomes that contain them.
- Inversion is an operator inspired by natural processes that has not yet been widely used.

REFERENCES

Goldberg, David E. (1989). *Genetic Algorithms in Search, Optimization and Machine Learning*. Reading, Mass.: Addison-Wesley.

Holland, John H. (1975). *Adaptation in Natural and Artificial Systems.* Ann Arbor: The University of Michigan Press.

Schaffer, J. David, Richard A. Caruana, Larry J. Eshelman, and Rajarshi Das (1989). A study of control parameters affecting online performance of genetic algorithms for function optimization. In J. David Schaffer (ed.), *Proceedings of the Third International Conference on Genetic Algorithms.* San Mateo, Calif.: Morgan Kaufmann Publishers.

2

Performance
Enhancements

A RUN OF GA 1-1

It is time to run GA 1-1. If you have loaded OOGA, enter a Lisp Listener and run the function (trial-run 'ga-1-1). A series of format statements describing the run will appear on your computer terminal. If you don't have OOGA or are unable to run it, no matter. The most important parts of the computer display are shown below. (Note: the output has been edited slightly to fit the page here. You do not see what the actual output shows, that OOGA used double-precision numbers here.)

AT 100 BEST 5 CHROMOSOMES ARE:

```
1000000001010001101110011100110101011011110110 .99026249
0111001100001010000110100000101100100011011 0 .98930211
1001100001111101101001100111100110101000111 0 .90970485
100111110000101101000010101110001011110000000 .86966422
101101101100010111100001110111110111101001000 .82411554
```

AT 200 BEST 5 CHROMOSOMES ARE:

```
100001111100000111100001110111110111101001001 .99229899
0111000011010111100111100001000100101101101 0 .98491267
011100110000101000011010001111001101110110 10 .97578980
011010000010001101100001111000100011010010 00 .96230820
011010000010001001010110010001101011101110 00 .94706181
```

AT 400 BEST 5 CHROMOSOMES ARE:

```
1000011111000011100110100011110011011101101 0 .98227694
1000011111001010000110100011110011011101101 0 .98225310
100001111100000100010110010001101011110000 0 .97738784
0111001100001010000110100011110011011101101 0 .97578980
011100110000101000011010001111001110110110 10 .97576120
```

    [...output deleted...]

AT 3000 BEST 5 CHROMOSOMES ARE:

```
011101111001000011010110001111110011100000 01 .98052087
0111001111110001010111100011110010111001001 1 .97721386
011100111111000001010110001111001011100100 11 .97720801
011100111111001011010010001111010011100000 11 .97699464
011100111111000001010110001111010011100000 11 .97698057
```

    [...output deleted...]

AT 4000 BEST 5 CHROMOSOMES ARE:

```
01111001011000101101011000100001100100010011 .99304112
01111011111000101101111000101000110110010010 .99261288
01111011110000100101011000101000110110010011 .99254826
01111011110000000101011000101000110110010001 .99254438
01111011111000101101011000101001110110010011 .99229856
```

We can make a number of observations here. Note that OOGA has output information about the state of the population after each generation has been produced. This information includes a listing of the top five chromosomes in the population of 100 chromosomes, as well as the evaluation given these five chromosomes by binary f6.

Note that the best solution generated after 200 individuals had been generated was not present after 400 individuals had been generated. The genetic material in that individual probably found its way into the population, but the individual itself was replaced without making any literal copies, and the evaluation of the best member of the population is worse than in previous generations.

Also observe that the top five chromosomes in the first generation are quite diverse. Since the population was initialized randomly, this is to be expected. At the end of generation 2—after 200 individuals have been created—the population is still fairly diverse. After 3000 individuals have been created, we see that many of the positions (the first five positions, for example) have the same value on each of the top five chromosomes. After 4000 individuals have been created, the top five chromosomes are more similar.

These effects are fairly typical of genetic algorithm runs. Here is some terminology to describe them. In the initial phase, the population is *heterogeneous* in that it consists of randomly-generated chromosomes that are quite dissimilar. At this point, the performance of the genetic algorithm is equivalent to that of an algorithm that generates random chromosomes and evaluates them. Beginning with generation 2, however, crossover begins to play a strong role in recombining features of the best chromosomes. This recombination phase lasts until 3000 individuals or so have been produced. At this point, the population has converged somewhat on a single solution. From then until the final generation, the genetic algorithm is searching around the region of that solution, with mutation the principal introducer of diversity. When a population consists primarily of similar individuals, we say that it has *converged*.

Finally, note that when the population has converged on a chromosome that would require mutation of a good many bits to cause any improvement, the run of the genetic algorithm is for all intents and purposes completed. Rather than continuing to run it to no avail, a better usage of computer cycles might be to run the algorithm again using a different random number seed and take the best result, or to use a hill-climbing heuristic to search methodically for improvements. This situation had not occurred when the present run ended, but it frequently occurs in runs of the genetic algorithms discussed later in this text.

We will be examining the performance of many genetic algorithms in this tutorial. A computer output of the sort given for GA 1-1 is not the best medium for displaying such performance. Instead, we will be using *performance graphs*. The performance graphs in this tutorial were generated as follows: A genetic algorithm was run 20 times. The performance of a single genetic algorithm at any point in the run is taken to be the evaluation of the best member of its current population. The performance graph is an average of these evaluations over the 20 runs. Since genetic algorithms are stochastic, their performance usually varies from run to run, and so a curve showing average performance is a more useful way to view the behavior of a genetic algorithm than a representation of the behavior of a genetic algorithm in a single run.

Figure 2.1 contains performance graphs for each of the algorithms we will be examining in this chapter. For GA 1-1, the genetic algorithm will be run for 40 generations of 100 individuals. Thus, the x-axis of the graph, running from 0 to 4000, is a measure of how many chromosomes have been created and evaluated at that point in the run. The y-axis displays the evaluation of the best individual in the population at that point in the run, but the meaning of the y-axis measurement requires some explanation.

The optimal value of the binary f6 function is 1.0, although given the 44-bit representation used here and double-precision arithmetic, the best solution possible is of the form 0.999..., where there are eight 9's after the decimal point. At least one of the genetic algorithms to come generates this solution frequently. If we plot the performance of our algorithms on this simple problem directly, it will be very difficult to see the difference between an optimal solution and 0.99304112, which is the best value produced by GA 1-1 in the printout reproduced earlier. Accordingly, what is graphed on the y-axis of the performance graphs here is *the number of nines immediately after the decimal point in the evaluation of the best population member*. The best our genetic algorithms can do with this measure is to produce eight nines after the decimal point. So the y-values on our graphs range from zero to eight. This measure is, in effect, a logarithmically sensitive indication of the precision of the solutions. For binary f6, if an algorithm finds the top of a hill adjacent to the optimal hill in the search

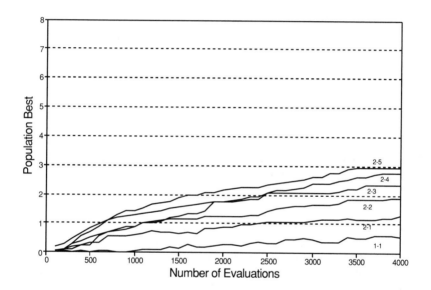

Figure 2.1: Performance Graphs for GA 1-1, GA 2-1, GA 2-2, GA 2-3, GA 2-4, and GA 2-5

space, the evaluation contains two nines after the decimal point and the y-value of this solution is 2. As you can see, this measure places a heavy premium on finding the optimal hill and on climbing to its very top.

The lowest curve in Figure 2.1 is the performance curve for GA 1-1 on binary f6. The other curves are for genetic algorithms yet to be described in this chapter. This curve for GA 2-5 is a characteristic one. As you can see, it rises rapidly at the beginning of the run, rises more slowly as the system nears a locally optimal solution, and flattens at the end. The shape of these curves is typical; with a genetic algorithm, as is the case for most optimization algorithms, most of the improvements come at the beginning of the run, and tiny increments in performance or no improvements tend to come at the end.

NOTES ON OOGA

You can skip this section if you have no interest in understanding the computerized version of OOGA.

Randomization in OOGA

A genetic algorithm will tend to display different performance every time it runs, as long as the random number generator has not been reset. If you call the function (`trial-run 'ga-1-1`) in OOGA, you may see different results from those I obtained. This can happen when we use different random number generators, or when we use the same generator initialized differently. For these reasons I have reported the average of multiple runs of each of the genetic algorithms here.

The Efficiency of OOGA

The methods used in OOGA are not designed for optimal efficiency. When a conflict has arisen here between efficiency and clarity of exposition, I have generally favored clarity. When producing systems for clients I use a larger and more complicated system that is more efficient but less clear. The primary intent here is to provide you with as clear an example of a working genetic algorithm as I can.

Finding Out More About OOGA

There are three ways to find out more about OOGA after you have read the documentation and run the tutorial examples. The first is to look at the components of the OOGA systems that are used as the examples in each chapter. The techniques incorporated into the demonstration genetic algorithm become more complicated and more powerful as the tutorial proceeds. A good way to understand the complicated techniques is to understand their simple counterparts, for the overall strategy in this tutorial is to show that tailoring a genetic algorithm to a real-world domain can be viewed as replacing the generic components of the three basic modules with more specialized ones.

A second way to learn about the architecture of the OOGA genetic algorithms is to type (display *ga*) after running any of the genetic algorithms. The result will be a screen display of the components and parameters of the algorithm you have just run.

A third way in which you can find out about OOGA involves tinkering with it. Genetic algorithm practitioners tend to present genetic algorithms logically and straightforwardly, but none of them learned about genetic algorithms logically or straightforwardly. Genetic algorithms are complicated, nonlinear systems with strongly ecological behavior. Making a single change in the system can affect the performance of each of its parts in ways that even expert practitioners cannot predict. At present, genetic algorithms are as much an art as a science, and art is learned through experience. If you have been thinking during this chapter about the way the genetic algorithm has been made to simulate natural evolution, you have probably had ideas about how to make the genetic algorithm work better. Some of those ideas will have been tried before. Some of them will not. You will learn things about the way genetic algorithms work that you cannot learn in other ways if you test your ideas empirically. And you may well learn things that will add to the body of knowledge in the field.

You might begin by investigating questions like "What happens if the crossover rate is increased?" "What happens if the mutation rate is increased?" "What happens if crossover is eliminated?" "What happens if mutation is eliminated?" You can investigate many such questions by changing a single line or word of OOGA code. Try it.

A note of advice is in order here. Remember that, since a genetic algorithm rarely repeats its performance on a complicated problem, you will need multiple runs to determine that a variation on the algorithm performs well on a problem. A single run could lead to a lucky hit or an improbable miss that gives misleading information about the hypothesis you are testing.

IMPROVING ON GA 1-1

In the remainder of this chapter we investigate four improvements to the performance of GA 1-1. In this and subsequent chapters, improvements are made by replacing techniques in the genetic algorithm without changing the overall architecture. The genetic algorithm that will result when all four changes are made is displayed below. The components of GA 1-1 that are replaced to make GA 2-5 are indicated with triple asterisks. Algorithms GA 2-1 through GA 2-4 are incremental stages in the transition from GA 2-1 to GA 2-5.

```
    -----------------------------------------------------------

    ANATOMY OF GA-2-5

    -----------------------------------------------------------

    EVALUATION-MODULE

        EVALUATION FUNCTION:  BINARY-F6

    -----------------------------------------------------------

    POPULATION-MODULE

        REPRESENTATION TECHNIQUE:  BINARY-REPRESENTATION
            Bit string length: 44
        INITIALIZATION TECHNIQUE:  RANDOM-BINARY-INITIALIZATION
***     DELETION TECHNIQUE: DELETE-LAST
***     REPRODUCTION TECHNIQUE:
***         STEADY-STATE-WITHOUT-DUPLICATES
        PARENT SELECTION TECHNIQUE:
            ROULETTE-WHEEL-PARENT-SELECTION
***     FITNESS TECHNIQUE: LINEAR-NORMALIZATION
***         INTERPOLATE FITNESS FROM 100 TO 1 BY 1
```

```
        PARAMETERIZATION TECHNIQUES: none


        Population size: 100
        Desired Trials: 4000
        Current Index: 0
```

```
    REPRODUCTION-MODULE

        OPERATOR SELECTION TECHNIQUE:   USE-FIRST-OPERATOR
        OPERATORS:
            ONE-POINT-CROSSOVER-AND-MUTATE
***            bit mutation rate: 0.04
***            crossover rate: 0.8

        PARAMETERIZATION TECHNIQUES: none

        Operator Weights: (100)
```

LINEAR NORMALIZATION

GA 1-1 finds better solutions to binary f6 than random search, but the performance of GA 1-1 will degrade if trivial changes are made to the function to be optimized. As an illustration, let's vary our evaluation function slightly, by changing the constant that is subtracted so that it is 999.5 instead of 0.5. Let's call the resultant function *elevated f6*. Here is the formula for elevated f6:

$$999.5 - \frac{(\sin \sqrt{x^2 + y^2})^2 - 0.5}{(1.0 + 0.001(x^2 + y^2))^2}$$

The curve determined by this formula looks exactly like that for binary f6 except that it is raised 999 units higher. The best chromosome for elevated f6 will be the best chromosome for this problem. The only difference is that the evaluation of each chromosome is increased by 999.

The effect this has on GA 1-1 is to make the average fitness curve flat over the 4000 individuals produced by the algorithm. The performance of GA 1-1 on what is essentially the same problem has been substantially degraded. Why is this?

The explanation has to do with the way GA 1-1 implements the notion of natural selection. How many offspring should we expect a chromosome in one generation to have in the next generation? For GA 1-1, the expected number of offspring for a chromosome equals that chromosome's evaluation divided by the average of the evaluations of all the chromosomes in the population. If a chromosome has an evaluation that is two times higher than that of the population average, we expect that it will contribute two offspring to the next generation. When GA 1-1 has initialized its population, the evaluation of the best chromosome in the population averages .979, the evaluation of the worst averages .066, and the average evaluation is .514. There is strong selection pressure here in favor of the best chromosome and pressure against the worst one.

When GA 1-1 initializes its population on elevated f6 with identical chromosomes, the evaluation of the best chromosome will be 999.979, the evaluation of the worst will be 999.066, and the average of the population will be 999.514. The best and the worst chromosomes will produce nearly identical numbers of offspring in the next population, and so the effect of natural selection on this problem is nearly nonexistent.

Genetic algorithm practitioners have solved this and related problems by converting the evaluations of the chromosomes into fitness values in various ways. I call techniques for performing this function *fitness techniques*, and we will replace evaluation is fitness in the Population Module of the genetic algorithm with one of them. Most of the techniques for generating fitnesses are not affected by the difference between f6 and elevated f6. Two fitness techniques that are commonly used are the following:

- *Windowing.* Find the minimum evaluation in the population. Assign each chromosome a fitness equal to the amount that it exceeds this minimum. Optionally, a minimum amount greater than the minimum value may be created as a guard against the lowest chromosomes having no chance at reproduction. Windowing is the default fitness technique in John Grefenstette's GENESIS system. You should read Grefenstette's chapter in Part II for a description of a more substantial version of the windowing algorithm.

Examples of Fitness Techniques

Original Evaluation	200	9	8	8	4	1
Fitness is Evaluation	200	9	8	8	4	1
Windowing with Minimum = 0	199	8	7	7	3	0
Windowing with Minimum = 10	199	10	10	10	10	10
Linear Fitness with Decrement = 1	100	99	98	97	96	95
Linear Fitness with Decrement = 20	100	80	60	40	20	1

Figure 2.2: Examples of the operation of three fitness techniques on a list of six chromosome fitnesses. The windowing technique is shown with two settings of the minimum fitness parameter. The linear fitness technique is shown with two settings of the decrement parameter and with a minimum value of 1.

- *Linear Normalization.* Order the chromosomes by decreasing evaluation. Create fitnesses that begin with a constant value and decrease linearly. The constant value and the rate of decrement are parameters of this technique. (Linear Normalization is the fitness technique that will be used in the remainder of this tutorial.)

Examples of these two normalization techniques and evaluation is fitness are found in Figure 2.2 where they are applied to a set of six fitnesses. These two techniques are illustrated using two parameter settings.

I have constructed the data in Figure 2.2 so that two common features of genetic algorithm runs are exemplified. The first is the existence of a chromosome that is much better than its nearest competitor. This chromosome—a *super individual*—will probably eliminate all its competitors in a generation or two when fitness is equal to evaluation. If this happens, the super individual will have had very few chances to try recombination with the other population members, and the population will probably experience rapid convergence. The second feature is that there are several individuals with evaluations that are very close. Genetic algorithms often find sets of close competitors, especially at the end of their runs. When there are close competitors, we may well want to increase the amount of selection pressure so that the best competitors reproduce disproportionately often.

Let's examine what the three normalization techniques do with these evaluations. Evaluation is fitness makes no alterations to the evaluations, and so the super individual and the close race persist. With windowing, both the super individual phenomenon and the close race persist but their effects are somewhat mitigated. Linear normalization selects for the super individual, but not so much that it will quickly dominate the population. This technique also heightens the competition in the close race. Note that the value of the decrement parameter in linear normalization is very important. The greater it is, the greater the selection pressure in favor of the highest rated chromosome.

Since linear normalization does better with both difficult phenomena, we will use it henceforth in this tutorial. GA 2-1 is constructed from GA 1-1 by replacing evaluation is fitness with linear normalization. Figure 2.1 compares the average performance of GA 1-1 and GA 2-1 on f6. You can see that GA 2-1 outperforms GA 1-1 significantly.

We have already discussed why GA 2-1 would do better on elevated f6. Why does it also do better on binary f6? The answer lies in reexamining the output for GA 1-1 reproduced earlier. As you can see, in the final generations the chromosomes in the population are extremely similar, as are their evaluations. Thus, the later phases of these runs display the close race phenomenon. GA 1-1 treats close competitors essentially as equals whereas GA 2-1 heightens the pressure on them, effectively giving reproduction chances to the ones with a slight edge that are far in excess of the amount by which they are superior.

The question of what technique should be used to convert evaluations to fitnesses is discussed in great detail in a dissertation by James Baker. For a summary of his findings, as well as a deeper discussion of parent selection techniques, see Baker (1989). (Baker's preferred parent selection technique—probably the best for generational replacement systems—has not been discussed here because we will no longer be using generational replacement.)

ELITISM

We noted earlier that the best member of the population may fail to produce offspring in the next generation. The *elitist* strategy fixes this potential source of loss by copying the best member of each generation into the succeeding generation. The elitist strategy may increase the speed of domination of a population by a super individual, but on balance it appears to improve genetic algorithm performance.

GA 2-2 is derived from GA 2-1 by replacing the generational replacement reproduction technique with generational replacement with elitism. As you can see in Figure 2.1, this causes an improvement in performance.

THE TRADITIONAL GENETIC ALGORITHM

Genetic algorithms vary from practitioner to practitioner, but it may be useful for you to know that GA 2-2 can be regarded as a *traditional genetic algorithm* in that much of the early work in the field (and some of the current work) is carried out using genetic algorithms that are very similar to GA 2-2. In subsequent chapters of this tutorial, when I refer to a "traditional genetic algorithm," I mean algorithms similar to GA 2-2. These algorithms use binary encodings, generational reproduction (usually with elitism), some fitness normalization technique, and the one-point crossover and mutate operator. The fitness technique, the parameters of the operators, and the parent selection techniques used may vary, but the performance of GA 2-2 will probably be similar to that of other traditional genetic algorithms using these techniques.

STEADY-STATE REPRODUCTION

When GA 2-2 reproduces, it replaces its entire set of parents by their children. This generational replacement technique has some potential drawbacks. One is that even with an elitist strategy, many of the best individuals found may not reproduce at all, and their genes may be lost. It is also possible that mutation or crossover may alter the best chromosomes' genes so that whatever was good about them is destroyed. Neither of these outcomes is desirable.

One solution to this problem is to modify the reproduction technique so that it replaces only one or two individuals at a time rather than all the individuals in the population. I believe this technique was first described in print in Whitley (1988), as part of Darrell Whitley's genetic algorithm called GENITOR. It has also been described in Syswerda (1989), where Syswerda called it *steady-state reproduction*—the name we will use here. The algorithm is described in Figure 2.3.

Steady-state reproduction has one parameter—the number of new chromosomes to create. Note that generational replacement is the special case of steady-state reproduction in which this parameter equals the size of the population. Practitioners of steady-state reproduction, however, typically create and insert just one or two children at a time.

Some genetic algorithm practitioners have tried the steady-state reproduction technique and found it to be inferior to generational replacement. To test this observation, let's build GA 2-3, a genetic algorithm that re-

---

Steady-State Reproduction

1. **Create n children through reproduction**
2. **Delete n members of the population to make space for the children**
3. **Evaluate and insert the children into the population**

Figure 2.3: The steady-state reproduction algorithm

sembles GA 2-1 except for two features. The generational-replacement reproduction technique in GA 2-1 is replaced by a steady-state reproduction technique, and the replacement technique becomes delete last. Delete last makes space for n new children by deleting the n worst members of the population.

Figure 2.1 shows the average performance of GA 2-3 on f6. As you can see, GA 2-3 does better than GA 2-1. How, then, could researchers have found it to be inferior? The answer to this question lies in the way I have graphed the average performance curves. The graph in Figure 2.1 has an x-axis measuring how many individuals have been created over the course of the run. GA 1-1, with a .60 crossover rate and a .008 mutation rate, will produce many individuals that are identical. So the number of individuals produced and the number of *distinct* individuals produced are different. Many researchers ignore duplicate chromosomes in their performance statistics. When they do so, the relative performance of GA 2-1 and GA 2-3 is reversed.

If we measure performance in this way, then, it appears that the steady-state reproduction technique is inferior to generational replacement. There is more to the story, however.

STEADY-STATE WITHOUT DUPLICATES

We have discussed a measure of performance by which steady-state reproduction is inferior to generational replacement. However, the steady-state technique makes possible another technique that leads to improved performance. This technique, which I call *steady-state without duplicates*, has been used by Whitley and Syswerda with great success. Steady-state without duplicates is a reproduction technique that discards children that are duplicates of current chromosomes in the population rather than inserting them into the population. When we use this reproduction technique, every member of the population will be different.

It is not obvious how to use this technique with generational replacement, and I do not know of anybody who has done so. For discussion of related ideas, however, you may wish to study the work of David Goldberg and his students on sharing (see Goldberg and Richardson (1987) and Deb and Goldberg (1989)). Those techniques lie off the main line of development here. You should read about them if you can, if only to learn different ways in which the genetic algorithm can be modified to improve its performance and reduce population convergence.

Steady-state without duplicates has one tremendous benefit. The genetic algorithms we have considered up to this point create their allotted number of chromosomes with a great deal of duplication. During the run of GA 1-1, for example, about half the chromosomes created were duplicates. Steady-state without duplicates allows much more efficient use of our allotted number of chromosomes by guaranteeing that reproduction never creates duplication in the population.

This technique involves some overhead, in that it requires the Population Module to carry out what is potentially a large number of equality tests whenever a child is created. However, this book is about applying genetic algorithms to real-world problems, and my experience in real-world domains is that most of the time spent by a real-world genetic algorithm is spent in evaluation.

The import of this fact is that the additional execution time spent by a real-world genetic algorithm using steady-state without duplicates is negligible with regard to the total time spent in optimization. Since this technique leads to performance benefits that offset its additional computation demands, it is worthwhile to use it when appropriate. This increase in computation time is significant when compared to the evaluation of binary f6, but binary f6 is not a real-world problem.

GA 2-4 is like GA 2-3 except that it replaces the steady-state reproduction technique with steady-state without duplicates. Figure 2.1 shows the performance curve for GA 2-4. (Note that we don't need elevated f6 any more now that we are using linear normalization in each of the algorithms we are comparing.) The performance curve for GA 2-4 is better than that for GA 2-3. (It is also better than that for GA 2-2, even using the nonduplication performance measure discussed above.)

We can make an important point in regard to these performance comparisons. As we saw in the preceding section, adding a simple steady-state reproduction technique to the genetic algorithm did cause a degradation in performance when we used a performance measure that ignores duplicate chromosomes. Another change to the genetic algorithm was necessary in order for the potential of steady-state reproduction to manifest itself. This is not a rare occurrence. Again and again researchers using genetic algorithms that differ in a number of ways are unable to reproduce each others' claims about a single technique by adding it to their own system and trying it out. Genetic algorithms are not simple systems; they are complicated and nonlinear, and their behavior is difficult to formalize and predict. As is the case in nature, a mutation of a genetic algorithm may require changes in others of its parts in order for its worth to be demonstrated.

IMPROVED PARAMETER VALUES

This section introduces you to some considerations having to do with the parameter values used by the components of genetic algorithms. A fact that you should know if you do not use the traditional genetic algorithm on your problems is that modifying the parameter settings or the components of a genetic algorithm may greatly impact the algorithm's performance.

The transition from GA 1-1 to GA 2-4 is an example of this phenomenon. GA 1-1 uses values for the parameters of crossover rate in one-point crossover and mutation rate in bit mutation that have been found to be near-optimal across a variety of problems. Those values are low since generational replacement must be conservative about destroying the genetic matter in its best individuals while building the next generation.

GA 2-3 and GA 2-4, however, build in elitism for nearly all the best population members, and so applying more crossover and a good deal more mutation will not cause any good chromosomes to be lost. Accordingly, it is possible to improve the performance of GA 2-4 by building GA 2-5, just like GA 2-4 except that the crossover rate parameter associated with the operator increases from .6 to .8, and the bit mutation rate increases from .008 to .04—a fivefold increase.

Figure 2.1 shows averaged performance curves for GA 2-5 on binary f6. As you can see, the new parameter values improve the performance of the genetic algorithm over those of the other genetic algorithms already discussed in this chapter.

BUILDING A NEW GENETIC ALGORITHM MAY NOT BE EASY

Let us return to a point made above. Getting from GA 2-1 to GA 2-5 required four improvements. If we had tried to get from GA 2-1 to GA 2-5 by making those improvements in a different order, we might have turned back, because some of these improvements, implemented singly, degrade the performance of the algorithm or leave it essentially unchanged. The researchers who began to work with genetic algorithms like GA 2-5 did not do so by working incrementally. They had an idea—in this case, the idea that a steady-state reproduction technique could outperform generational replacement. They worked on a genetic algorithm containing that component, accepting the initial poor performance while they tinkered with whatever components of the algorithm they could. In the end, they even altered parameter values that were regarded as very good ones by researchers using generational replacement.

The result of this work was a better genetic algorithm for solving problems of the type we are considering. Deriving this genetic algorithm, however, was a process similar to that in nature in which new mutations occur in populations isolated from the main population. In such cases, members of the population may find adaptations to the mutation, finally yielding a coadapted set of features that works better than do the features of the main population. If these mutations had arisen in the main population, their initial decrease in fitness would probably have led to their being eliminated by the pressures of natural selection.

ON ROBUSTNESS

This is a good time to introduce the notion of *robustness*, a theme that we will consider several times. Steady-state without duplicates does not work well when the evaluation function is *noisy*—that is, when the evaluation function may return different values each time the same chromosome is evaluated.

Many real-world problems display just such nondeterministic behavior. One example is that of finding good parameter settings for a genetic algorithm by using a genetic algorithm. Each chromosome in the population contains a list of parameter values. Chromosomes are evaluated by installing those parameter values in a second genetic algorithm that does a trial run of the problem to be solved. Since multiple runs of the second genetic algorithm will produce different results even with identical parameter settings, a chromosome that is a list of parameter values has no single evaluation, and this can cause complications for a genetic algorithm using steady-state without duplicates. The reason is that a lucky good evaluation for a chromosome will be forever associated with it, and, since a good chromosome won't be deleted from the population using this technique, the spurious evaluation will cause the bad genetic material to disseminate throughout the population.

This problem doesn't occur when we use the generational replacement technique, in which a chromosome is reevaluated whenever it turns up in the population. Generational replacement works for noisy evaluation functions, and it also works for deterministic ones. This fact implies that generational replacement has an important property that genetic algorithm researchers have pursued. An algorithm is *robust* insofar as it displays good performance across a variety of problem types. Given a variety of problem types that are well weighted with noisy evaluation functions, we would expect to see generational replacement with its parameter settings outperform steady-state without duplicates with its parameter settings. And so generational replacement is more robust than steady-state without duplicates.

Developing a robust form of the genetic algorithm has been an early and continuing topic of research in the genetic algorithm community. It has sometimes been presented as the principal goal of genetic algorithm research. My own point of view on this matter is quite different. My goal is to show that genetic algorithms are the *best* optimization algorithms on certain classes of problems. This goal is in general incompatible with the goal of producing a robust genetic algorithm, because, in general, the robustness of a genetic algorithm and its performance on a particular problem are inversely related.

I mean by this that in order to derive a high-performance genetic algorithm on a domain, we are often required to specialize the genetic algorithm for the domain. Such specialization may entail the use of techniques adapted especially for the type of problem being solved, parameter settings adapted to those techniques, operators that are specific to the problem domain, and so on. But such specialization generally entails loss of robustness. In the case of GA 2-3, GA 2-4, and GA 2-5, we have gained performance on a problem that has a deterministic evaluation function, but we have lost performance on the full class of problems that we might encounter. In my view it is fine to gain performance based on our *knowledge* that the evaluation function is deterministic. Researchers whose goal is robustness want instead to produce a single algorithm that can be applied to deterministic and noisy problems.

Although no one has yet published a paper about steady-state reproduction with noisy evaluation functions, it seems possible that modifications to the steady-state algorithm can be devised that will outperform generational replacement on such problems. This is a fertile topic for future research.

SUMMARY

The principal points addressed in this chapter were the following:

- Some technique for converting evaluations into fitness is needed in order for genetic algorithm performance to be robust. We use linear normalization here.

- Elitism enhances the performance of a genetic algorithm using generational replacement.

- Steady-state reproduction without duplicates can improve the performance of a genetic algorithm on deterministic problems.

- Using these reproduction techniques, we can improve performance by increasing the crossover and mutation rates.

- Steady-state reproduction may not work well when the evaluation function is noisy. This implies that steady-state reproduction is less robust than generational replacement unless the steady-state reproduction technique is modified in some way to accommodate noise.

REFERENCES

Baker, J. E. (1989). Reducing bias and inefficiency in the selection algorithm. In J. David Schaffer (ed.), *Proceedings of the Third International Conference on Genetic Algorithms*. San Mateo, Calif.: Morgan Kaufmann Publishers.

Deb, Kalyanmoy, and David E. Goldberg. (1989). An investigation of niche and species formation in genetic function optimization. In J. David Schaffer (ed.), *Proceedings of the Third International Conference on Genetic Algorithms*. San Mateo, Calif.: Morgan Kaufmann Publishers.

Goldberg, David E., and J. Richardson. (1987). Genetic algorithms with sharing for multimodal function optimization. In John J. Grefenstette (ed.), *Genetic Algorithms and Their Applications: Proceedings of the Second International*

Conference on Genetic Algorithms. Hillsdale, N.J.: Lawrence Erlbaum Associates.

Syswerda, Gilbert. (1989). Uniform crossover in genetic algorithms. In J. David Schaffer (ed.), *Proceedings of the Third International Conference on Genetic Algorithms.* San Mateo, Calif.: Morgan Kaufmann Publishers.

Whitley, Darrell. (1988). GENITOR: a different genetic algorithm. In *Proceedings of the Rocky Mountain Conference on Artificial Intelligence.* Denver, Colo.

3
Further Evolution of the Genetic Algorithm

In this chapter we consider three modifications to GA 2-5 that further improve its performance. The anatomy of the final result is displayed below, with triple asterisks in the left margin indicating the modifications described in this chapter:

```
--------------------------------------------------------------

ANATOMY OF GA-3-4

--------------------------------------------------------------

EVALUATION-MODULE

    EVALUATION FUNCTION:  BINARY-F6

--------------------------------------------------------------
```

```
STEADY-STATE-POPULATION-MODULE

    REPRESENTATION TECHNIQUE:   BINARY-REPRESENTATION
        Bit string length: 44
    INITIALIZATION TECHNIQUE:   RANDOM-BINARY-INITIALIZATION
    DELETION TECHNIQUE:  DELETE-LAST
    REPRODUCTION TECHNIQUE:
        STEADY-STATE-WITHOUT-DUPLICATES
    PARENT SELECTION TECHNIQUE:
        ROULETTE-WHEEL-PARENT-SELECTION
    FITNESS TECHNIQUE:  LINEAR-NORMALIZATION
        INTERPOLATE FITNESS
        FROM 100 TO 1 BY 1
    PARAMETERIZATION TECHNIQUES:
***        INTERPOLATE FITNESS DECREMENT
***        FROM 0.2 to 1.2

    Population size: 100
    Desired Trials: 4000
    Current Index: 0
```

```
REPRODUCTION-MODULE-3-1

    OPERATOR SELECTION TECHNIQUE:
        ROULETTE-WHEEL-OPERATOR-SELECTION
    OPERATORS:
***        UNIFORM-LIST-CROSSOVER
***        BINARY-MUTATION
        bit mutation rate: 0.04
    PARAMETERIZATION TECHNIQUES:
***        INTERPOLATE OPERATOR WEIGHTS
***        FROM (70 30) to (50 50)

***    Operator Weights: (70 30)
```

The following sections describe these modifications.

INDEPENDENT OPERATORS

So far we have used only one operator—one-point crossover and mutate—in our genetic algorithms. A good deal of research in the genetic algorithm field has been devoted to investigating the effects of variations on these operators, as well as to developing new kinds of operators tailored to different encoding schemes and specific problem types. This research often entails adding other operators to the operator list in the Reproduction Module. From this point on, we will be modifying the operator list. Let us begin by splitting the one-point crossover and mutate operator into its two components.

The one-point crossover and mutate operator is really two operators—a one-point crossover operator and a binary mutation operator—bundled into one. From now on, these and other operators will be handled separately, as follows.

First, the one-point crossover and mutate operator will be replaced by two operators: one-point crossover and binary mutation. One or the other of these operators will be applied during a reproduction event, but not both. In particular, mutation will no longer be applied to the result of crossover—a possibility with the combined operator which, in my view, confuses things.

Second, reproduction will be *operator-based*. Operator-based reproduction occurs when each reproduction event consists of the selection of an operator from the operator list, the application of the operator to an appropriate number of parents, and the return of the children produced to the Population Module. Since we have only had one operator in the genetic algorithms we have been considering, we have been using this sort of reproduction by default. The principal modification here is one in orientation, reflected in the changes discussed next.

Third, each operator has a new parameter, its fitness. For simplicity's sake, I will always manipulate the fitness parameters of the operators in the operator list so that they total 100. Given more than one operator, we will now see operator fitness lists in the Reproduction Module that contain several numbers whose total is 100.

Finally, selection of operators for reproduction is now roulette wheel operator selection. This means that when we are to select an operator for a reproduction event, we apply the roulette wheel technique already being used to choose parents. Since the operator fitnesses always total 100, an operator's percentage chance of being selected for a reproduction event will equal its fitness.

These changes introduce some theoretical complications into a genetic algorithm. Operator-based reproduction carried out in this way entails that the number of children an operator is likely to have is a function of its fitness, but that function depends on how many children each of the operators produces. In the case of a crossover operator with a fitness of 60 and a mutation operator with a fitness of 40, we see that the crossover operator has a 60 percent chance of being chosen for any reproduction event and the mutation operator has a 40 percent chance. Let's assume that the crossover operator produces two children and that the mutation operator produces one. Thus, in a genetic algorithm allowing duplicate population members, these fitnesses entail that the crossover operator will actually produce 75 percent of the children while the mutation operator will produce only 25 percent.

This complication in computing the true reproduction rate of operators is compounded by another that we have already considered. When no duplicates are allowed in the population, it is impossible to say with certainty what percentage of children each operator is really producing. If the population is quite similar and a crossover operator is producing individuals that are already represented in the population, the percentage of children it produces may be lower than these figures suggest. Similarly, if a mutation operator is applied at a low rate so that some children have no mutated fields, or so that a a single mutation is likely to create a chromosome that is already in the population, then in a genetic algorithm eliminating duplicate population members the children of the mutation operator will not be inserted into the population. The number of children produced by the mutation operator will also be lower than its fitness would suggest.

These features of the genetic algorithms we are considering can be troublesome, since we now have a system for which we can no longer say with certainty how the operators will perform. Some of the modifications described later—especially the adaptive operator fitness technique discussed in Chapter 7—can ease these troubles.

Modifying GA 2-5 in the three ways described above creates GA 3-1. The performance curve for GA 3-1 is found in Figure 3.1 and is seen to be quite similar to that of GA 2-5.

OTHER CROSSOVERS

Now that our operators have been unbundled, let's examine our crossover operator. Although one-point crossover was inspired by biological processes, its algorithmic counterpart has some drawbacks. One of the most

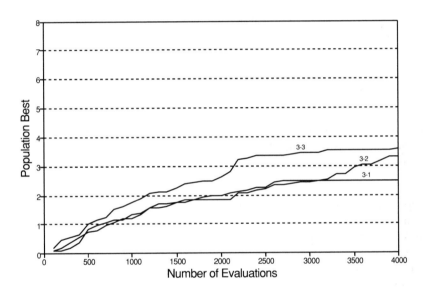

Figure 3.1: Performance curves for GA 3-1, GA 3-2, and GA 3-3 on binary f6

Two Schemata that One-Point Crossover Cannot Combine

Chromosome 1	**1**	1	**0**	1	1	0	0	1	0	1	1	0	1	**1**
Chromosome 2	0	0	0	1	**0**	1	1	0	**1**	**1**	**1**	**1**	0	0

Figure 3.2: Example of two schemata that one-point crossover cannot combine. The emphasized bits comprise the schemata.

Two-Point Crossover Can Combine the Schemata

Parent 1	**1**	1	**0**	1	1	0	0	1	0	1	**1**	**0**	**1**	**1**
Parent 2	0	0	0	1	**0**	1	1	0	**1**	**1**	1	1	0	0
Child 1	**1**	1	**0**	1	**0**	1	1	0	**1**	**1**	**1**	**0**	**1**	**1**
Child 2	0	0	0	1	1	0	0	1	0	1	1	1	0	0

Figure 3.3: The schemata can be combined with two-point crossover. The emphasized bits comprise the schemata. Vertical lines indicate the two crossover points.

important is that one-point crossover cannot combine certain combinations of features encoded on chromosomes. Consider the two chromosomes in Figure 3.2, where the indicated bits represent two high-performance schemata. Let's assume that if either of these schemata is altered, its beneficial effects are lost. One-point crossover cannot cause these schemata to be combined on a single chromosome because the first schema is encoded in bits at both ends of the first chromosome. No matter where the crossover point is selected, the first schema will be broken up and will not be passed on.

Until recently, the most popular solution for this problem has been the use of two-point crossover. This operator is like one-point crossover, except that two cut points rather than one are selected at random, and chromosomal material is swapped between the two cut points. Figure 3.3 shows two-point crossover applied to the chromosomes in Figure 3.2. With two-point crossover the problematic schemata are combined in one of the children.

However, there are schemata that two-point crossover cannot combine. This fact has caused genetic algorithm researchers to consider other crossover operators with more cut points. The operator we will use here is described in Syswerda (1989) and has the capability to combine any schemata that do not disagree at any single position. Syswerda calls this

An Example of Uniform Crossover

Parent 1	1	0	0	1	0	1	1
Parent 2	**0**	**1**	**0**	**1**	**1**	**0**	**1**
Template	1	1	0	1	0	0	1

Yields

Child 1	1	0	**0**	1	1	**0**	1
Child 2	**0**	**1**	0	**1**	0	1	**1**

Figure 3.4: An example of uniform crossover. Bits on the second parent are enclosed in boxes. The template determines which child acquires these bits.

operator *uniform crossover*. It works as follows. Two parents are selected, and two children are produced. For each bit position on the two children, we decide randomly which parent contributes its bit value to which child. Figure 3.4 shows uniform crossover in action. For each bit position on the parents, the template indicates which parent will contribute its value in that position to the first child. The second child receives the bit value in that position from the other parent.

Uniform crossover differs from one-point and two-point crossover in several ways. First, the location of the encoding of a feature on a chromosome is irrelevant to uniform crossover. With one-point or two-point crossover, the more bits that intervene in a schema, the less likely it is that the schema will be passed on to a child intact. (A potentially dangerous feature of uniform crossover, however, is that it can do a great deal of violence to whatever is good on a chromosome.) Second, one-point and two-point crossover operators are more local than uniform crossover and are more likely to preserve good features that are encoded compactly. Syswerda has carried out experiments showing that for some problems the ability of uniform crossover to combine features no matter where they are located outweighs the global devastation it is possible to wreak when using this operator on radically dissimilar chromosomes.

Some genetic algorithm researchers have tried similar experiments and have found uniform crossover to be inferior to two-point crossover. However, the genetic algorithms employed in their experiments used generational replacement and allowed duplicate population members, factors that may account for the variation in results. You can see the difference for binary f6 by running GA 3-2. The performance curve for GA 3-2 is slightly better than that of GA 3-1. The only difference between the two algorithms is the replacement of two-point crossover on the operator list of GA 3-1 with the uniform crossover operator.

INTERPOLATING PARAMETER VALUES

The *fixed parameters* of a genetic algorithm are those of its parameters that do not alter over the course of a run. So far all the parameters in the genetic algorithms we have used have been fixed. We can gain improvements in performance, however, if we alter some of these parameters as the run is in progress. In this section we investigate the effects of interpolating parameter settings in GA 3-2 in order to produce GA 3-3.

Interpolating Operator Fitness

We will create GA 3-3 by altering the relative fitnesses of our operators over the course of the run. In order to do so, we will put a technique on the list of parameterization techniques in the Reproduction Module. This technique—interpolate operator weights—has as parameters the beginning and end values of the operator fitness list. Over the course of a run this technique will cause the value of the operator fitness list to be set to the beginning value, to change gradually over the run, and to be set finally to the end value when the run is completed.

Why would it be a good idea to alter the fitnesses of the various operators? The answer is an interesting one. There are strong interactions among the relative weights of operators in a genetic algorithm, the degree of convergence of the population, and the nearness of the best individuals to the local solution optima. At any point in a run there is a ratio of operator fitnesses that will lead to best results, and this optimal ratio changes throughout the run.

This fact may seem counterintuitive because of the following argument. Suppose someone claims that the optimal set of operator fitnesses at a certain point in a run gives 80 percent fitness to crossover and 20 percent fitness to mutation. In such a case, the argument goes, why won't we get better results by giving the dominant operator—crossover—100 percent of the computation cycles and forgetting about mutation altogether?

The answer to this question can be found by experiment. If you delete either mutation or crossover from the operator set in GA 3-2, you will find that the performance of the system is impaired. The operators support each other in important ways. Mutation introduces diversity into the population, and crossover acts to combine the good schemata already present. If mutation is deleted, the population converges rapidly with some important schemata missing. Once every bit at a given position has converged, crossover can never introduce diversity at the position. If crossover is deleted, the system uses mutation to search the area in the neighborhood of a number of points without the capability of combining the best features of those points, and this leads to diminished performance.

Thus a judicious blend of mutation and crossover operators does better than either one alone. Another important point is that the optimal ratio changes as the population changes over the course of a run. In my experience, interpolating the fitness values of a genetic algorithm's operators over the course of a run always does better than leaving them fixed. At worst, a good interpolation will do exactly as well, since fixed values are a trivial case of interpolated ones. But I have never seen a problem of any complexity for which the optimal fitnesses of multiple operators were level over the course of the run.

Figure 3.1 shows the average performance curve of GA 3-3. GA 3-3 is like GA 3-2 except that it interpolates the uniform crossover operator fitnesses from 70 to 50 and the bit mutation fitness from 30 to 50 over the course of the run. The performance of the genetic algorithm improves when the fitness values are changed in this way, although the improvement is not dramatic. (It is much more dramatic for algorithms with more operators, such as those in Chapter 5).

Why do these relative fitnesses change? The answer to this question probably has to do with the way the population changes over the course of a run of GA 3-3. After initialization, the population is diverse. Scattered throughout it are pieces of good solutions to the problem. Crossover is the best way to put those pieces together, and this accounts for the large value of the uniform crossover fitness at the beginning of the run. Once these pieces have been assembled, however, the population has converged in a small section of the search space where crossover is of much less use. Now mutation is the operator that must be used to search for better solutions in the region of the best current solutions. This accounts for the increase in binary mutation fitness as the run proceeds.

You may wonder where I got these fitness values. They were not pulled out of a hat. In Chapter 7 we will discuss an adaptive parameterization technique that produced them efficiently and effectively.

AN OOGA NOTE

Although I presented the interpolation components above as linear algorithms, in OOGA the interpolation components don't actually carry out interpolation after every reproductive event. For efficiency's sake, interpolation is done only after every 50 new individuals (the number is a parameter of the algorithm) have been inserted into the population. Thus, the actual fitness parameter settings in the OOGA implementation look like a series of stairs on a staircase. These flat lines approximate the linear values, however, and so in this case efficiency has won out over clarity in the construction of OOGA.

SUMMARY

The principal points addressed in this chapter were the following:

- Distinguishing the various genetic operators can lead to performance improvements and can clarify our understanding of each operator's effects.

- When operators are distinct, operator-based reproduction is a useful reproduction technique.

- When there is more than one operator, operators can compete to carry out reproduction, just as parents compete.

- When there is more than one operator and no duplication is allowed, computing the true reproduction rate of each operator can be difficult.

- Uniform crossover and two-point crossover are attractive alternatives to one-point crossover for the problems we are considering.

- Interpolating genetic algorithms can achieve better performance than algorithms with fixed parameters, but in order to use interpolation techniques we must determine the value of additional parameters.

- The optimal ratio of operator fitnesses changes over the course of a run of a genetic algorithm.

REFERENCES

Syswerda, Gilbert (1989). Uniform crossover in genetic algorithms. In J. David Schaffer (ed.), *Proceedings of the Third International Conference on Genetic Algorithms*. San Mateo, Calif.: Morgan Kaufmann Publishers.

4
Hybrid Genetic Algorithms

In this chapter I will describe a way to create genetic algorithms that do well on optimization problems for which there are existing algorithms, optimization heuristics, or domain knowledge that can aid in optimization. This technique will often yield an optimization algorithm that outperforms its competitors. It does so by incorporating what is best in its competitors into the genetic algorithm itself. We can call the result of this incorporation process a *hybrid* genetic algorithm.

WHY HYBRIDIZE?

A central goal of the research effort in genetic algorithms is to find a form of the algorithm that is *robust*—that performs well across a variety of problem types. This goal has been well served by the use of binary representation and a single, generic operator. The binary representation can encode nearly anything, and the operator does not include any knowledge about the domains on which the optimization may be taking place.

Those of us who optimize for a living have a quite different goal. In order for us to be paid to carry out a genetic algorithm application, we must persuade persons with a difficult optimization problem and the funds to solve it that we will create the *best* algorithm for solving their problem, given their computing and time constraints. Although genetic algorithms using binary representation and single-point crossover and binary mutation are robust algorithms, they are almost never the *best* algorithms to use for any problem. It is a sad fact of life for us, as it is in nature, that a species of individuals that do well across a variety of environmental niches but are never the best in any niche must give rise to a variety of better adapted species or fail in the competition for resources.

Genetic algorithm practitioners who optimize for a living must do what nature does: we must adapt to fit our intended niches. For us, the intended niches are the hard, real-world optimization problems we are trying to solve. If this adaptation process does not yield the most fit algorithm in those niches, we will fail because, as is true in nature, there is intense competition for resources in the world of industry. People with real-world problems do not want to pay for multiple solutions to be produced at great expense and then compared. They prefer instead to concentrate their funds on development of the algorithm likely to do best on their problem.

Before a project to apply genetic algorithms to an optimization problem is approved, genetic algorithm practitioners are often required to predict the performance of their techniques as compared with traditional techniques developed over time on the problem domain; artificial intelligence techniques that may look promising compared to a blind optimizer; machine learning techniques that have also been tailored to the domain; and any of the variety of new or enhanced techniques that continually arise in the computer science and optimization fields and that inspire intense loyalty on the part of their adherents.

Successful predictions of this kind cannot be made for the genetic algorithms we have investigated so far. In fact, many genetic algorithm researchers are not interested in discussions of this type because they already know that their generic algorithms will not compete successfully with approaches specialized for a problem. That is not their goal.

It is through discussions of this type, however, that industrial genetic algorithm practitioners obtain resources. Over the last seven years I have used three techniques that lead to success when these discussions take place. With these techniques I am regularly able to convince people with difficult optimization problems that I can build an optimization algorithm for them that will outperform that of its competitors. What is more, by applying these techniques I have created algorithms that perform as predicted. In the next section I describe these three techniques so that you can apply them too.

HOW TO ADAPT A GENETIC ALGORITHM

Let us call the person whose real-world problem you are proposing to solve the *user*. If the user knows the problem well, there is probably an algorithm for solving it that the user understands and whose performance, though not optimal, is known. Let us call this algorithm the *current algorithm*. The user will probably be familiar with the current algorithm and the way it solves the problem. Unless you can persuade the user that your way of solving the problem will perform at least as well as the current algorithm, the user will have little reason to request your services. Furthermore, if you are proposing to use an encoding technique like the bit string one that is unfamiliar to the user and whose use seems peculiar for the problem at hand, your chances of obtaining resources from the user to implement your solution to the problem may be decreased.

When I talk to the user, I explain that my plan is to hybridize the genetic algorithm technique and the current algorithm by employing the following three principles:

Use the Current Encoding. Use the current algorithm's encoding technique in the hybrid algorithm.

Hybridize Where Possible. Incorporate the positive features of the current algorithm in the hybrid algorithm.

Adapt the Genetic Operators. Create crossover and mutation operators for the new type of encoding by analogy with bit string crossover and mutation operators. Incorporate domain-based heuristics as operators as well.

A project employing these principles can be very easy to explain and sell to the user. I discuss each principle and its implications below.

Use the Current Encoding

Using the current encoding has two advantages. First, it guarantees that the domain expertise embodied in the encoding used by the current algorithm will be preserved. Second, it guarantees that the hybrid genetic algorithm will feel natural to the user, since the hybrid algorithm will be operating on the same kind of structures that the current algorithm is working on.

Using the current encoding is the way I have produced the encoding techniques I have used to solve many real-world problems. It is an effective way to generate powerful optimization algorithms.

Hybridize Where Possible

This principle tells us to incorporate whatever optimization techniques the current algorithm uses into the hybrid genetic algorithm. This can be done in a variety of ways, including these:

- If the current algorithm is a quick one, the hybrid algorithm can add the solution or solutions that it produces to the initial population. In this way a hybrid genetic algorithm with elitism is guaranteed to do no worse than the current algorithm does. In general, crossing the solutions of the current algorithm over with each other or with other solutions will lead to improvements.

- If the current algorithm carries out successive transformations on encodings, it can be very useful to incorporate those transformations into your operator set. For example, in Montana and Davis (1989) we describe the performance of a hybrid genetic algorithm that uses *backpropagation*—a neural network training technique—as an operator, together with mutation and crossover operators that are tailored to the neural network domain. This hybrid algorithm was developed by following the three principles we are discussing.

- If the current algorithm is clever in interpreting its encodings, it can be important to incorporate whatever it does into the decoding part of the evaluation technique. Algorithms adapted to a problem domain often contain coadapted encoding and decoding strategies. We have already borrowed the encoding strategy. This principle may entail borrowing features of the current decoding strategy also.

Adapt the Genetic Operators

The two preceding principles tell us to incorporate what is good about the current algorithm into the hybrid algorithm. This principle tells us to incorporate what is good about the genetic algorithm as well.

Note that once we have adopted the current algorithm's encoding strategy, we can no longer apply the familiar genetic algorithm operators that manipulate bit strings. Now we must create their analogs given the encoding technique that has been adopted. This is not always easy to do.

Crossover operators, viewed in the abstract, are operators that combine subparts of two parent chromosomes to produce new children. The adopted encoding technique should support operators of this type, but it is up to you to combine your understanding of the problem, the encoding technique, and the function of crossover in order to figure out what those operators will be. In succeeding chapters of this tutorial I will give you some examples of ways to do this. If you can't create useful crossover operators for your problem based on your encoding technique, you should probably look for another encoding technique or concede the field to the current algorithm, for it is the mechanism of crossover that generally leads to improved performance over the current algorithm.

The situation is similar for mutation operators. We have decided to use an encoding technique that is tailored to the problem domain; the creators of the current algorithm have done this tailoring for us. Viewed in the abstract, a mutation operator is an operator that introduces variations into the chromosome. As you will see later in this chapter these variations can be global or local, but they are critical to keeping the genetic pot boiling. You will have to combine your knowledge of the problem, the encoding technique, and the function of mutation in a genetic algorithm to develop one or more mutation operators for the problem domain. Subsequent chapters of this tutorial contain multiple examples of such operators. Again, if there do not seem to be useful mutation operators given the encoding technique that has been adopted, then the encoding technique should be changed or a genetic algorithm approach may not be the right approach to the problem at hand.

Finally, there may be rules of thumb that are known to work sometimes when solving problems in the domain. Such heuristics are often fallible, but they lead to improved solutions frequently enough to be of use. It can be extremely useful to incorporate such heuristics in the operator set so that they can be triggered randomly like the other operators, and so that they provide some domain-based guidance to the search process.

GENERAL COMMENTS ON HYBRIDIZATION

I use the term *hybrid genetic algorithm* for algorithms created by applying these three principles. The connotations of this term are very useful. When I propose a hybridization project to the user and to the proponents of the current algorithm, the terminology makes it clear that the current algorithm will be used in solving the problem. This removes a potential source of friction, especially since the user is often the originator of the current algorithm or has been extensively involved in its development for the problem domain.

The employment of these three principles allows us to make a very compelling case to the user during discussions about the approach to solving the problem. If the work to be done includes the application of these three principles, the resultant hybrid algorithm will likely do better than a robust genetic algorithm, since the hybrid system will incorporate a good deal of existing algorithmic expertise about the domain. It will almost certainly do better than the current algorithm, since the good features of the current algorithm are part of the hybrid algorithm, as are the genetic algorithm techniques of population management and crossover.

When I propose to use a hybrid system, the user generally responds favorably, especially if the response is based on the technical merits of the various alternatives. In general I have found that the user's response is appropriate. Hybridizing a genetic algorithm with the current algorithm tends to yield hybrid vigor, in that the hybrid child will outperform each of its parents on the problem domain.

Note that during the preceding discussion I have assumed that only one current algorithm is available on the problem domain. When multiple current algorithms are available, these principles should be adapted appropriately. Obvious adaptations are to hybridize only the best techniques from the current algorithms, to incorporate mechanisms from each as operators, and to build selected techniques from them into the decoder.

SUMMARY

The principal points addressed in this chapter were the following:

- Traditional genetic algorithms, though robust, are generally not the most successful optimization algorithm on any particular domain.

- Hybridizing a genetic algorithm with algorithms currently in use can produce an algorithm better than the genetic algorithm and the current algorithms.

- Hybridization entails using the representation and optimization techniques already in use in the domain, while tailoring the genetic algorithm operators to the new representation.

- Hybridization can entail adding domain-based optimization heuristics to the genetic algorithm's operator set.

- Hybridization techniques have generated a number of successful optimization algorithms.

REFERENCES

Montana, David J., and Lawrence Davis. (1987). Training feedforward neural networks using genetic algorithms. *Proceedings of the 1989 International Joint Conference on Artificial Intelligence.* San Mateo, Calif.: Morgan Kaufmann Publishers.

5

Hybridization and Numerical Representation

In this chapter we apply the hybridization techniques described in Chapter 4 to an algorithm for optimizing f6. The result is GA 5-1, an algorithm that is a hybrid of GA 3-3 and the f6 optimization algorithm. This hybrid algorithm will outperform its parents.

A RANDOM GENERATE AND TEST ALGORITHM FOR F6

Figure 5.1 describes a technique for finding solutions to f6. For obvious reasons this technique is called *random generate and test*.

Random generate and test is not the most effective technique in the domain, but it will hybridize with our genetic algorithm surprisingly well. Like the traditional genetic algorithm it is quite robust. We can apply a random generate and test strategy to a wide variety of optimization problems, and it sometimes does surprisingly well.

The Random Generate and Test Algorithm

1. Randomly generate a pair of numbers within the allowable limits for x and y.

2. Evaluate f6 for this pair of values, with the first value used as x in f6 and the second value used as y.

3. Store the x and y values away with their evaluation.

4. If time has not run out, go to 1. Otherwise, return the stored x and y values that have the best evaluation.

Figure 5.1: The random generate and test algorithm.

In what follows we go through the steps of hybridizing random generate and test with the genetic algorithm that was developed in Chapter 3. The result will be GA 5-1.

USING THE CURRENT ENCODING

The random generate and test algorithm is a simple one, which makes it a good algorithm to hybridize for our first example. Algorithms of this type are also frequently used as a basis of comparison for other optimization algorithms. Its main point of interest for us here is that it does not represent its solutions in binary notation. Instead, as is true of nearly every numerical function optimization method in common use, its solutions are represented as lists of real numbers. This point merits some discussion.

Why Replace Bit String Encoding?

Bit string encoding is the most common encoding technique used by genetic algorithm researchers. Consequently, many people have used this technique in carrying out real-world applications of genetic algorithms, with varying results.

Bit strings have several advantages over other encodings. They are simple to create and manipulate. They are theoretically tractable, in that their simplicity makes it easy to prove theorems about them. Performance theorems have been proved for bit string chromosomes that demonstrate the power of natural selection on bit string encodings. Just about anything can be encoded in bit strings, so one-point crossover and mutate, using the robust parameter settings of a robust genetic algorithm, can be applied without modification to a wide range of problems. Thus, some people in the genetic algorithm field have come to believe that bit strings should be used as the encoding technique whenever they apply a genetic algorithm.

My experience doesn't support this view. At the time of this writing, I have done nine real-world applications of genetic algorithms across a variety of problem types. The encoding techniques used in those applications have varied widely, but none used bit string encoding. In fact, when I carried out performance tests on those domains, a genetic algorithm using bit string representation was always outperformed by the encoding techniques I ultimately used.

The previous chapter of this tutorial described three techniques for deriving encoding techniques, operators, and associated components of a genetic algorithm that will make it the best available optimization algorithm for a specific problem domain. The first technique involved using the representation already in use in the field. Practitioners of other types of optimization algorithms do not tend to use bit string encodings. Using this hybridization technique implies, then, that bit string encodings will probably not be employed in hybrid algorithms.

This feature of hybrid genetic algorithms elicits varying reactions from researchers in the genetic algorithm field. One's feelings for and against binary encoding can be very strong. Some researchers refer to binary genetic algorithms as "real" genetic algorithms and leave unspoken their characterizations of the rest. This situation will probably change as researchers prove theorems for other encoding techniques like the theorems that have been proved for the binary encoding technique, and as robust operator sets and parameter settings are settled on for other types of encodings. (During the final stages of preparation of this book, David Goldberg has completed a preliminary draft of a paper proving performance theorems for genetic algorithms with real number representation. Goldberg's paper may prove to have a bridging effect when discussions about representation issues occur in the genetic algorithm community.)

An underlying reason for differences in feeling about bit string encoding may well be that researchers studying the performance of genetic algorithms tend to have different goals from those of us who are applying genetic algorithms to real problems. The researchers sometimes characterize themselves as attempting to find a robust algorithm that does well across the widest variety of problem types; this is the characterization set forth by David Goldberg (1989). Goldberg avoids adding any domain knowledge or representational specialization to his genetic algorithms, preferring instead to develop a robust "black box" problem solver. This goal is a useful one in its own right, and work toward it has to some degree helped set the stage for real-world applications. But it is a goal orthogonal to that of producing the best optimization algorithms for particular problems.

Adopting Real Number Representation

We will alter our genetic algorithm so that it uses real numbers in its chromosomes. The alterations required to the Population Module are as follows:

- We replace the binary encoding technique of GA 3-3 with a real number list technique. (This change has little practical import. It just documents the type of structure that the genetic algorithm will be working on.)

- We replace the binary f6 evaluation function of GA 3-3 with real number f6, an evaluation function that is just like binary f6 except that no conversion from binary to real number notation is required.

- We replace the random binary initialization technique of GA 3-3 with random real number initialization. This technique generates lists of real numbers that fall within limits that are parameters of the technique.

Here is a chromosome generated by the initialization method just described: (1563404, 11204).

This chromosome looks quite different from a chromosome using the bit string representation. From the human factors point of view, this way of encoding solutions to the problem presents certain advantages. For one thing, we humans are used to seeing numbers and can understand a real number encoding more easily than we can understand a string of bits that encodes a real number. But this advantage is an artifact of our having ten fingers. If we thought in base 2, perhaps binary encodings would be more natural to us. The real number encoding technique gives us some other advantages, however, that binary encoding does not. We consider them later in this chapter.

HYBRIDIZING WHERE POSSIBLE

There isn't a great deal to hybridize in the random generate and test algorithm. The genetic algorithms we have been using already employ random generation and testing to initialize their populations. After that, crossover is used to combine good features of the randomly generated chromosomes, and mutation is used to explore chromosomes that are similar. An interesting question that bears directly on the performance of genetic algorithms is, will a run with 100 randomly generated chromosomes that give rise to 3900 descendants through mutation and crossover do better than a run of 4000 randomly generated chromosomes? The answer is yes. The performance curve of an algorithm that randomly generates chromosomes lies almost exactly on that of GA 2-2.

ADAPTING THE GENETIC OPERATORS

Now we come to the interesting part—changes to the Reproduction Module. We are given a very simple encoding: a list of two real numbers. Our task is to figure out crossover and mutation operators for this encoding that are analogous to the binary crossover and mutation operators. Although most of the problems which genetic algorithm researchers have studied involve numerical function optimizing, the studies themselves have been carried out using the binary encoding technique. As a result, little has been published on operators for real number chromosomes. I expect this situation to change in the future. For now, let me suggest several obvious mutation and crossover operators. We will put them into the genetic algorithm and see how they work.

Crossover Operators for Real Number Encodings

Encoding solutions as a list of two real numbers does not rule out one-point crossover or uniform crossover. There was nothing about those crossover operators that relied on binary representation; all they require is lists. So we already have crossover operators that will work on this problem, and we will continue to include uniform crossover in our operator list.

But we can do more. We have available to us something we did not have before—a representation we know is numeric. GA 3-3 achieves robustness by making few assumptions about its subject matter. We are hybridizing a current algorithm that we know will be dealing with numbers, and so it is now possible to use crossover operators that incorporate a numerical flavor. I will add one at this point: average crossover. This crossover operator takes two parents and produces one child that is the result of averaging the corresponding fields of two parents. Once you begin to think about numerical crossover operators, many other candidates come to mind. We will use this one for now.

Mutation Operators for Real Number Encodings

What is a useful real number analog of genetic mutation? I have experimented with two possibilities.

The first simply replaces a real number in a chromosome with a randomly selected real number, if a probability test is passed. Let's call this operator real number mutation. Real number mutation functions in the spirit of the random generate and test algorithm. However, it is likely to apply to only one of the two real numbers on the chromosome, and so it has some chance of doing better than random generate and test, since half the chromosome will retain its old value and that value may be a useful one to hold constant during search.

The second and third operators are more clever. They are variations of an operator type called real number creep. The idea behind a creep operator is that a chromosome that is reproducing is probably in a fairly good place with respect to the rest of the population. The function we are optimizing is a continuous one with hills and valleys. If we are on a good hill, we want to jump around on it, to move nearer to the top. Real number creep can have that effect. What it does is sweep along the chromosome, creeping any value up or down a small and random amount if a probability test is passed. The maximum amount that this operator can alter the value of a field is a parameter of the operator. So is the probability of its altering any field. In GA 5-1, there are two operators of this type, differing only in the maximum amount that each can creep.

We can't creep to neighboring positions in the binary notation without converting the binary encoding into a real number, creeping, and converting back. True practitioners of the traditional genetic algorithm may avoid using such operators because they are not robust—they will do badly on non-numerical problems. The only researcher I know of who uses a creep-type operator with binary representation is Darrell Whitley, who reports that creeping single steps in binary notation on numerical problems can lead to successful results (Whitley 1988).

THE PERFORMANCE OF GA 5-1.

The result of these hybridization maneuvers is GA 5-1, a genetic algorithm whose architecture is displayed in OOGA as follows, where triple asterisks in the left column denote new components:

```
-----------------------------------------------------------------

ANATOMY OF GA-5-1

-----------------------------------------------------------------


    EVALUATION-MODULE-5-1

***    EVALUATION FUNCTION:   REAL-NUMBER-F6

-----------------------------------------------------------------


    POPULATION-MODULE-5-1

***    REPRESENTATION TECHNIQUE:   REAL-NUMBER-REPRESENTATION
***    INITIALIZATION TECHNIQUE:
***       RANDOM-REAL-NUMBER-INITIALIZATION
       DELETION TECHNIQUE:   DELETE-LAST
       REPRODUCTION TECHNIQUE:
          STEADY-STATE-WITHOUT-DUPLICATES
       PARENT SELECTION TECHNIQUE:
          ROULETTE-WHEEL-PARENT-SELECTION
       FITNESS TECHNIQUE:   LINEAR-NORMALIZATION
             INTERPOLATE FITNESS
             FROM 100 TO 1 BY 1
       PARAMETERIZATION TECHNIQUES:
             INTERPOLATE FITNESS DECREMENT
             FROM 0.2 to 1.23~

    Population size: 100
    Desired Trials: 4000
    Current Index: 0

-----------------------------------------------------------------
```

```
REPRODUCTION-MODULE-5-1

        OPERATOR SELECTION TECHNIQUE:
          ROULETTE-WHEEL-OPERATOR-SELECTION
        OPERATORS:
***       UNIFORM-LIST-CROSSOVER
***       AVERAGE-CROSSOVER
***       REAL-NUMBER-MUTATION
***         mutation rate: 0.5
***         mutation specs: ((0 4194303 T))
***       REAL-NUMBER-CREEP
***         creep rate: 0.7
***         creep specs: ((70000 T))
***       REAL-NUMBER-CREEP
***         creep rate: 0.7
***         creep specs: ((2000 T))
        PARAMETERIZATION TECHNIQUES:
          INTERPOLATE OPERATOR WEIGHTS
***             FROM (10 40 10 30 10)
***             TO (10 20 0 40 30)

    Operator Weights: (10 40 10 30 10)

----------------------------------------------------------------
```

Note the two creep operators. They differ only in the maximum amount of creep that they can cause. The first creeps a chromosome value an amount between 0 and 70,000. The second creeps a chromosome value an amount between 0 and 2000.

Figure 5.2 compares the performance of a new set of runs of GA 3-3 and a set of runs of GA 5-1 on f6. GA 5-1 was created by hybridizing a very simple optimization technique with the genetic algorithm strategy, and you can see that it does much better than the genetic algorithm it descended from. However, GA 5-1 is less robust than the genetic algorithms we have considered earlier.

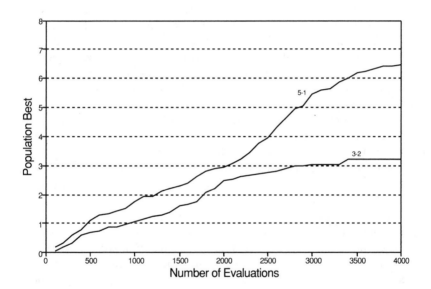

Figure 5.2: Performance graphs for GA 3-3 and GA 5-1 on f6.

CONCLUDING COMMENTS ON F6

This chapter concludes our use of f6 as a test problem. F6 is a simple problem in many ways, and so it has been a good environment in which to watch an elementary genetic algorithm evolve and adapt. GA 5-1, our final genetic algorithm, solves the problem better than domain-based algorithms and better than our more generic genetic algorithms, with concomitant loss of robustness.

If you have some experience with optimizing numerical functions, you may know an algorithm which will do better than these. That's good. I have not claimed to have produced the best algorithm for solving f6 using the three principles of hybridization I described in Chapter 4. I don't know any way to do that. What I have described is a set of techniques that may allow you to create an algorithm that will do better than any other algorithm on the problem, once you are told what that other algorithm is.

Indeed, if you know of a better algorithm for this problem, I recommend that you try hybridizing it with the genetic algorithm. My prediction is that your algorithm will be improved on hybridization. This sort of hybridization is just what I propose to do in discussions with the user. It does not matter a great deal in discussions with the user what the competing algorithms are on a problem domain, as long as we understand them and can figure out how to hybridize them with a genetic algorithm. Remember: in discussions with the user, as is the case in nature, you don't have to have the best possible solution to an environmental problem to compete for resources in the problem niche. You only need a solution that is better than any of the others you are competing with. Using nature's own techniques of mimicry and adaptation, as exemplified in the hybridization techniques we have discussed here, can help you to create a hybrid genetic algorithm that dominates in its niche.

SUMMARY

The principal points addressed in this chapter were the following:

- Numerical representation is effective on mathematical problems like f6.
- Numerical representation supports the use of numerical operators.
- Numerical operators may greatly improve the performance of the genetic algorithm on numerical problems.

REFERENCES

Goldberg, David E. (1989). *Genetic Algorithms in Search, Optimization and Machine Learning.* Reading, Mass.: Addison-Wesley.

Whitley, Darrell. (1988). GENITOR: a different genetic algorithm. In *Proceedings of the Rocky Mountain Conference on Artificial Intelligence.* Denver, Colo.

6

Order-Based Genetic Algorithms and the Graph Coloring Problem

INTRODUCTION

In this chapter we consider another type of optimization problem, the graph coloring problem. This problem is a combinatorial optimization problem rather than a numerical optimization problem, and the encoding technique used here will be quite different from the encoding techniques we have considered so far. The algorithm we hybridize in this chapter—a *greedy algorithm*—is an extremely common type. In this chapter you will learn a way to improve the performance of such algorithms by hybridization with a genetic algorithm wherever you encounter them.

THE GRAPH COLORING PROBLEM

A great many optimization problems are similar to the problem we consider in this section. The graph coloring problem concerns formal structures called *graphs*. A graph is a set of points called *nodes* with connections called *links* between pairs of nodes. Figure 6.1 shows an example of a graph. Put intuitively, we can make a graph as follows. Draw some points on a piece of paper. These are the graph's nodes. Draw a number of lines, each connecting a pair of nodes. These are the graph's links.

In Figure 6.1 each circle is a node. There are two numbers in each circle. The first is the node's *index*. These indices will be used to refer to the nodes. The second number is a *weight* assigned to the node for optimization purposes. The graph coloring problem concerns graphs like this, with weighted nodes. The problem itself is stated in Figure 6.2.

A simple graph coloring problem is as follows. Given the graph in Figure 6.1 and just one color, maximize the weight of the colored nodes.

Note that the nine nodes in the figure have different weights and that the links between them make it impossible to color every node. Nodes 7 and 9, for example, are linked. These nodes are the two with the highest weight, and at most one of them can be colored.

One strategy for the graph coloring problem that yields a promising solution in a single try is given in Figure 6.3. A *greedy algorithm* is an optimization algorithm that proceeds through a series of alternatives by making the best decision, as computed locally, at each point in the series. The algorithm in Figure 6.3 is a greedy algorithm because it colors the graph node by node, giving each node a color if it can and taking no account of the more global consequences of the colorings. The algorithm is also greedy because it orders its decisions so that what are likely to be the most important decisions are decided first and what are likely to be the least important are decided last.

The greedy node coloring algorithm is very quick because it considers only one of the possible colorings of each graph. Yet it is likely to do well with that one coloring, since it guarantees that the most heavily weighted nodes get colored first if possible. The algorithm will produce the optimal solution when applied to the graph in Figure 6.1. There the algorithm considers the nodes in the order (9 7 8 4 2 6 5 1 3). It colors node 9 and then leaves 7 uncolored because 7 is connected to 9 and 9 is colored. It leaves 8 uncolored for the same reason, colors 4, colors 2, and leaves the other nodes uncolored. Coloring nodes 9, 4, and 2 gives the algorithm a score of 15 + 9 + 8, or 32.

Let us consider a second graph, the one shown in Figure 6.4. This graph has seven nodes. The first six nodes are arranged in a ring around the outside of the graph, and the seventh is placed at the center of the ring, with links to every other node. The seventh node is also given the highest weight. The greedy algorithm will attempt to color the nodes in the order

Graph 1

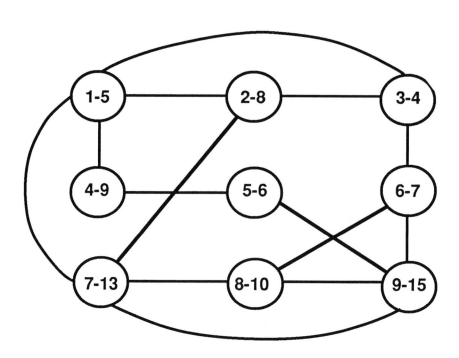

Figure 6.1: Example graph 1. Circles indicate nodes and lines indicate links. Inside each circle is the node's index and weight.

The Graph Coloring Problem

Given a graph with a weight on each node, and given n colors, the graph coloring problem is to achieve the highest score as follows:

1. Assign any of the nodes on the graph a color from the set of n colors if you like, but

2. No pair of nodes connected by a link can have the same color.

3. Your score is the total weight of the colored nodes.

Figure 6.2: The graph coloring problem

The Greedy Node Coloring Algorithm

1. Sort the set of nodes on the graph in order of decreasing weight.

2. Take each node in this sorted set in order and give it the first legal color it can have.

Figure 6.3: The greedy node coloring algorithm

Graph 2

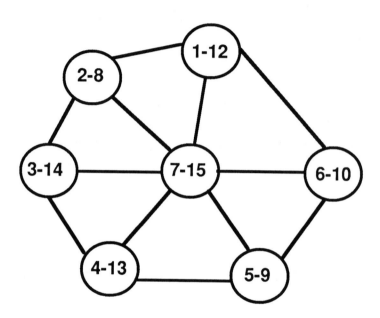

Figure 6.4: Example graph 2, a "hub and spokes" graph.

(7 3 4 1 6 5 2). If we are allowed only one color, the greedy algorithm will not do very well on this problem. Since the central node is the first node it considers, it will color the central node. However, now none of the other nodes can be colored, since they are all connected to the central node. This solution, though initially plausible, gets a score of 15, and that is not very good. The optimal solution is to color nodes 1, 5, and 3 to obtain a score of $12 + 9 + 14$, or 35. In this case, the greediness of the greedy algorithm has worked against it. Note that if we increased the weight of the central node in this graph so that it exceeded 35, then the greedy algorithm would find the optimal solution.

With two colors, the best solution for the first graph is to assign the first color to nodes 9, 4, and 2 and the second color to nodes 7, 6, and 5. The best solution for the second graph is to assign the first color to nodes 1, 5, and 3 and the second color to nodes 2, 4, and 6.

Altering the weights on the nodes and the number of colors available can dramatically alter the optimal solution for the graph coloring problem. Even minor variations on the problem can cause the optimal solution to change greatly. For these reasons this problem is not an easy one to solve. In the language of complexity theory, the graph coloring problem is NP-complete. Put crudely, this means that no strategy for solving the problem is guaranteed to succeed quickly and optimally for all possible graph coloring problems.

Because the greedy graph coloring algorithm is a good one for the amount of thinking it does, it seems promising to hybridize it with a genetic algorithm. But it is not immediately obvious how to do so. The greedy graph coloring algorithm does not seem to use any binary encoding, and there is no numerical encoding. How would such a hybridization proceed?

In the succeeding sections of this chapter we consider techniques that accomplish this hybridization. These techniques will work for a number of other greedy algorithms as well. The hybridization will be carried out by replacing various components in GA 3-4 to create GA 6-1. In the next section we discuss the techniques in GA 6-1 that differ from those in GA 3-4.

ORDER-BASED REPRESENTATIONS

The hybridization strategy outlined in Chapter 4 suggests that we use the encoding of our target algorithm for the hybrid genetic algorithm. The encoding technique used by the greedy algorithm is an *ordering* of the nodes in the graph. The greedy graph coloring algorithm creates an ordering based on descending weight and then decodes that ordering by giving each node in order the first legal color it can have based on the assignments of colors to previous nodes.

This technique for encoding solutions to the graph coloring problem—order the nodes on the graph in some way, and then decode the ordering just as the greedy algorithm did—is the one we will use for the hybrid genetic algorithm. Let's examine it in more detail.

The first change to GA 3-4 is to replace the binary f6 evaluation function with a node coloring evaluator. This evaluation function takes each of the nodes on a chromosome in order and assigns each the first legal color that node can have, if any. (Each assignment of a color takes into account the assignments already made.) The greedy algorithm applies the same technique to a list of nodes sorted by decreasing weight.

A *permutation* of a list of items is the result of scrambling the order of the items on the list. (A trivial permutation just retains the original order.) The next change to GA 3-4 is to replace the representation technique with a permuted list representation technique. This means that every chromosome used by the genetic algorithm will be a permutation of the list of nodes on the graph. The greedy algorithm we are hybridizing sorted the list of nodes and fed the result into a greedy decoder. We are going to use the genetic algorithm to search for better orderings of the list of nodes, and feed them into the same decoder.

The next change to GA 3-4 is to replace the initialization technique with random permutation. We will no longer generate real numbers or bit strings to serve as our encoding. Instead, the initialization technique will generate a random permutation of the list of nodes. Another feature of the greedy algorithm that we could also incorporate is the chromosome it generates. The greedy algorithm only generates a single chromosome, but the chromosome has a good chance of being better than average. We could "seed" our population with the chromosome that the greedy algorithm generates—the one whose order is based on the weights of the nodes. A "seeding" process guarantees that we can do no worse than the greedy algorithm we are hybridizing, since we are using its encoding as one of the members of our initial population and we are using an elitist population management technique. In the case of the problem we consider later this seeding process gains little, and so it is not used here, although seeding is a valuable technique for many problems.

More complicated changes are described below.

ORDER-BASED CROSSOVER

The most interesting parts of the hybridization process lie in the changes to the Reproduction Module, where we install operators that will work on the order-based representation. As we see below, these operators are quite different from those we have examined in earlier chapters. The greedy algorithm contributes no operators, so whatever operators we use must be derived by analogy from the operators we are familiar with.

Creating Crossover Operators for Order-Based Chromosomes

Some research has been done on crossover operators for order-based chromosomes (see Goldberg, 1989, and Oliver et al., 1987). I have obtained best results with a different crossover operator from any described in these papers. This operator—uniform order-based crossover—is the analog of uniform crossover, translated into the order-based realm. It is somewhat complicated. The way it works is detailed in Figure 6.5.

This operator does for order-based chromosomes what uniform crossover does for lists: it preserves part of the first parent while it incorporates information from the second parent. The information that is encoded here, however, is not a fixed value associated with a position on the chromosome. Rather, it is relative orderings of elements on the chromosomes. Parent 1 may have a number of elements ordered relatively well. Uniform order-based crossover allows Parent 2 to tell Parent 1 that others of its elements should be ordered differently. The net effect of uniform order-based crossover is to combine the relative orderings of nodes on the two parent chromosomes in the two children.

It is important to understand that what is being passed back and forth here is not information of the form "node 3 is in the fifth position." Instead, this operator combines information of the form "node 3 comes before node 2 and after node 7." The schemata in an order-based representation can be written in just this way. Let us denote the nodes in a permutation by their indices. Then the chromosome (5 1 6 4 2 3) contains a number of schemata, including (5 4), (1 4 2 3), (5 6 3), and (5 1 6 4 2 3).

An example of the action of this operator is found in Figure 6.6. As you can see, part of the structure of each parent is fixed by one parent (as determined by the bit string). The rest is reordered so that the elements in it have the same order they had on the other parent.

Uniform Order-Based Crossover

Given Parent 1 and Parent 2, create Child 1 in this way:

1. Generate a bit string that is the same length as the parents.

2. Fill in some of the positions on Child 1 by copying them from Parent 1 wherever the bit string contains a "1". (Now we have Child 1 filled in wherever the bit string contained a "1" and we have gaps wherever the bit string contained a "0".)

3. Make a list of the elements from Parent 1 associated with a "0" in the bit string.

4. Permute these elements so that they appear in the same order they appear in on Parent 2.

5. Fill these permuted elements in the gaps on Child 1 in the order generated in 4.

To make Child 2, carry out a similar process.

Figure 6.5: The uniform order-based crossover operator

Uniform Order-Based Crossover With:

Parent 1	1	2	3	4	5	6	7	8
Parent 2	8	6	4	2	7	5	3	1
Binary Template	0	1	1	0	1	1	0	0

Yields After Step 2:

Child 1	-	2	3	-	5	6	-	-
Child 2	8	-	-	2	-	-	3	1

Yields These Final Children:

Child 1	8	2	3	4	5	6	7	1
Child 2	8	4	5	2	6	7	3	1

Figure 6.6: An example of uniform order-based crossover.

An Order-Based Mutation Operator

Mutation operators carry out local modifications of chromosomes. The mutation operators we have considered previously have operated on one field of the chromosome at a time. We cannot operate on a single field of an order-based chromosome because there is nothing apparent to do to a single field. It is tempting to think of mutation as the swapping of the values of two fields on the chromosome. I have tried this on several different problems, however, and it doesn't work as well for me as an operator I call scramble sublist mutation. This operator selects a sublist of the items on a parent order-based chromosome and permutes them in the child, leaving the rest of the chromosome as it was in the parent. Scramble sublist mutation has no parameters, although a variation that I have found useful when chromosomes are quite long involves limiting the length of the section of the chromosome that can be scrambled. Figure 6.7 shows an example of scramble sublist mutation in operation.

Many other types of mutation can be employed on order-based problems. Scramble sublist mutation is the most general one I have used. To date nothing has been published on these types of operators, although this is a promising topic for future work.

Scramble Sublist Mutation

Parent   1   2   3 | 4   5   6   7   8 | 9

Child   1   2   3 | 6   4   8   7   5 | 9

Figure 6.7: Example of scramble sublist mutation. The elements between the two vertical lines are permuted.

THE ANATOMY OF GA 6-1

When all these changes have been made, we derive GA 6-1 from GA 3-4. The anatomy of GA 6-1 is as follows, where new components are noted with triple asterisks:

```
----------------------------------------------------------------

ANATOMY OF GA-6-1

----------------------------------------------------------------

EVALUATION-MODULE-6-1

    EVALUATION FUNCTION:  NODE-COLORING-EVALUATOR

----------------------------------------------------------------


POPULATION-MODULE-6-1

***   REPRESENTATION TECHNIQUE:  PERMUTED-LIST
***   INITIALIZATION TECHNIQUE:  RANDOM-PERMUTATION
      DELETION TECHNIQUE:  DELETE-LAST
      REPRODUCTION TECHNIQUE:
          STEADY-STATE-WITHOUT-DUPLICATES
      PARENT SELECTION TECHNIQUE:
          ROULETTE-WHEEL-PARENT-SELECTION
      FITNESS TECHNIQUE:  LINEAR-NORMALIZATION
```

```
           INTERPOLATE FITNESS
           FROM 100 TO 1 BY 1
        PARAMETERIZATION TECHNIQUES:
           INTERPOLATE FITNESS DECREMENT
           FROM 0.2 TO 1.2

        Population size: 100
        Desired Trials: 4000
        Current Index: 0

--------------------------------------------------------------

    REPRODUCTION-MODULE-6-1

        OPERATOR SELECTION TECHNIQUE:
           ROULETTE-WHEEL-OPERATOR-SELECTION
        OPERATORS:
***        UNIFORM-ORDER-BASED-CROSSOVER
***        SCRAMBLE-SUBLIST-MUTATION
        PARAMETERIZATION TECHNIQUES:
           INTERPOLATE OPERATOR WEIGHTS
***        FROM (60 40) TO (30 70)

        Operator Weights: (60 40)

--------------------------------------------------------------
```

A GRAPH COLORING PROBLEM

The following printout describes a 100-node graph coloring problem. The information is organized by nodes. Each of the hundred nodes is given an index, a weight, and a list of the nodes that it is connected to. For this problem, one is allowed three colors for coloring the graph. This problem is much more difficult than the two examples given earlier.

```
(1 62 (20 58 74 82))
(2 183 (6 12 20 28 29 32 51 53 56 70 79 84 94))
(3 247 (18 24 33 50 88 92))
```

(4 66 (70 74 75 79 95 98))
(5 181 (7 25 32 34 44 55 69 85))
(6 95 (2 62 67 84 91))
(7 112 (5 43 47 82 84))
(8 65 (10 20 25 71 72 91))
(9 163 (32 44 46 62 67 69 71 82 92))
(10 112 (8 34 40 43 76 83 88 93))
(11 153 (12 18 23 26 30 73 82 97))
(12 117 (2 11 16 17 25 31 36 44 69 71 72 80 84))
(13 163 (28 29 38 61 67 77 92))
(14 239 (25 33 61 92))
(15 193 (19 25 38 56 57 67 88 96 100))
(16 241 (12 25 40 42 64 68))
(17 255 (12 23 30 39 79 82))
(18 153 (3 11 36 58 59 73 80 90 96))
(19 191 (15 31 35 47 49))
(20 209 (1 2 8 31 61 73 100))
(21 97 (22 27 28 32 88 93))
(22 133 (21 52 63 71 82 89 94 100))
(23 84 (11 17 25 37 49 62 71 84 90 93))
(24 103 (3 26 43 55 56 58 66 72 98))
(25 81 (5 8 12 14 15 16 23 36 61 63 75 87))
(26 104 (11 24 37 41 46 53 64 68 94))
(27 220 (21 29 32 40 53 65 74 78))
(28 208 (2 13 21 42 68 72 79 87))
(29 187 (2 13 27 40 43 60 64 71 99 100))
(30 129 (11 17 52 54 60 67))
(31 65 (12 19 20 39 42 56 71 78 83 89 90 93))
(32 181 (2 5 9 21 27 35 37 38 49 50 68 73 79))
(33 141 (3 14 35 36 40 49 62 76))
(34 118 (5 10 36 41 55 87 100))
(35 81 (19 32 33 38 40 44 55 77))
(36 70 (12 18 25 33 34 46 50 53 70 78 81 91))
(37 210 (23 26 32 60 88 97))
(38 95 (13 15 32 35 50 60 61 78 88))
(39 103 (17 31 64 77))
(40 187 (10 16 27 29 33 35 51 53 82 86))
(41 121 (26 34 81 96))
(42 97 (16 28 31 51 56 75 76 78 87 94))
(43 130 (7 10 24 29 70))
(44 113 (5 9 12 35 70 74 81 91 100))

(45 169 (53 78 81 86))
(46 182 (9 26 36 50 54 59 63 83 92 96 98))
(47 232 (7 19 64 77 84 92))
(48 233 (49 84 88))
(49 250 (19 23 32 33 48 59 60 68 77 83 89 91))
(50 220 (3 32 36 38 46 55 57 84 86 87 97))
(51 117 (2 40 42 57 69 98))
(52 126 (22 30 61 81 84 99))
(53 84 (2 26 27 36 40 45 54 55 93 97 99))
(54 182 (30 46 53 57 58 69 95))
(55 145 (5 24 34 35 50 53 79 87))
(56 176 (2 15 24 31 42 67 71 89 92))
(57 241 (15 50 51 54 62 65))
(58 178 (1 18 24 54 59 67 79 88))
(59 226 (18 46 49 58 64 82))
(60 242 (29 30 37 38 49 62 82 90 91 100))
(61 153 (13 14 20 25 38 52 70 77 86))
(62 79 (6 9 23 33 57 60 63 77 88))
(63 236 (22 25 46 62 68 72 85 94 98))
(64 106 (16 26 29 39 47 59 76 85 96))
(65 218 (27 57 82 96))
(66 205 (24 67 84 96 97))
(67 154 (6 9 13 15 30 56 58 66 76 99))
(68 221 (16 26 28 32 49 63 69 79))
(69 164 (5 9 12 51 54 68 79 89))
(70 104 (2 4 36 43 44 61 77))
(71 105 (8 9 12 22 23 29 31 56 88 95))
(72 212 (8 12 24 28 63 86 87 97))
(73 218 (11 18 20 32 84 85 93 97))
(74 90 (1 4 27 44 77 88 92 95))
(75 193 (4 25 42 81 99))
(76 242 (10 33 42 64 67 78 85 86))
(77 236 (13 35 39 47 49 61 62 70 74 80 93 95))
(78 86 (27 31 36 38 42 45 76 93))
(79 118 (2 4 17 28 32 55 58 68 69 96))
(80 72 (12 18 77 94 99))
(81 234 (36 41 44 45 52 75 97))
(82 125 (1 7 9 11 17 22 40 59 60 65 88 94 98))
(83 90 (10 31 46 49))
(84 153 (2 6 7 12 23 47 48 50 52 66 73 93 95 99))
(85 199 (5 63 64 73 76 88))

(86 154 (40 45 50 61 72 76 87 89 91 93))
(87 107 (25 28 34 42 50 55 72 86 100))
(88 79 (3 10 15 21 37 38 48 58 62 71 74 82 85))
(89 75 (22 31 49 56 69 86 96 99))
(90 76 (18 23 31 60))
(91 229 (6 8 36 44 49 60 86))
(92 182 (3 9 13 14 46 47 56 74))
(93 251 (10 21 23 31 53 73 77 78 84 86 97))
(94 250 (2 22 26 42 63 80 82 97))
(95 85 (4 54 71 74 77 84))
(96 174 (15 18 41 46 64 65 66 79 89))
(97 219 (11 37 50 53 66 72 73 81 93 94))
(98 100 (4 24 46 51 63 82))
(99 254 (29 52 53 67 75 80 84 89))
(100 145 (15 20 22 29 34 44 60 87))

Figure 6.8 shows average performance curves for GA 6-1 on this problem, compared with an average performance curve for a random permute and test algorithm that proceeds by generating 4000 random permutations of the list of nodes and feeding them into the decoder. Random permute and test does not do as well as the hybrid genetic algorithm over the same number of evaluations.

One of the solutions found by the genetic algorithm during my preparation of this chapter gains a score of 10,594. This solution appeared in a genetic algorithm with a population size of 1200 and 10,000 desired trials. There are probably better solutions to be found, but solutions at this level do not tend to appear in populations of 100 chromosomes running for 4000 trials.

The greedy algorithm produces a solution with a score of 9,590. The hybrid genetic algorithm does better than the greedy algorithm it is based on. In addition, its best chromosome after 4000 individuals have been evaluated is about 7% better than the best individual generated by an equivalent number of random permutations. (The amount by which the genetic algorithm does better than random generate and test will vary as the problem varies.)

This simple order-based genetic algorithm with just two operators is fairly close to the system used in trials at U S West described by Cox, Qiu, and me in Part II of this book. The principal differences are in the population sizes and the fitness technique—there we use an exponentially decreasing rather than linearly decreasing fitness technique.

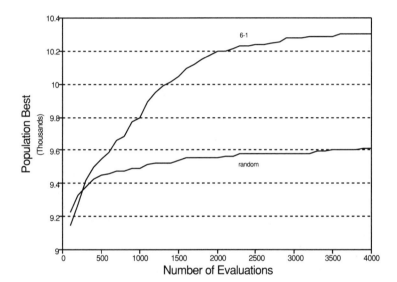

Figure 6.8: Performance curves for GA 6-1 and random generation on a 3-color graph coloring problem. The greedy algorithm scores 9,590.

IS THE ORDER-BASED ENCODING NECESSARY?

The graph coloring problem could be solved using another type of encoding strategy that is more akin to the encoding strategies of earlier chapters. This strategy would involve associating a node with each field of the chromosome and generating chromosomes by putting color values in each field.

This encoding strategy encodes the colorings of the graph literally. There are several problems with it and several possible solutions to those problems. I say "possible" because nobody has ever used literal encodings on the graph coloring problem. The principal problem with a literal encoding is that all the crossover operators we considered earlier—one-point crossover, two-point crossover, and uniform crossover, as well as a mutation operator that randomly replaces the color in a field of the chromosome—will tend to generate illegal solutions when they are applied. Two different colorings of a graph that are crossed over will tend to generate linked nodes that are given the same color, and such solutions are illegal.

A useful rule of thumb in selecting algorithms for optimization problems is this: an algorithm that generates many illegal solutions will perform worse than one that generates no illegal solutions. Unless something is done, a genetic algorithm that uses the operators above will fall prey to this generalization. Note that GA 6-1 can never generate an illegal solution to the graph coloring problem because the decoder never assigns an illegal color. This is an advantage of the order-based decoding strategy over the literal strategy, unless something else is done.

A technique that has been used to good effect when using literal representations on other problems (the "traveling salesman problem" is the most common) is to apply operators of the sort we have just considered and then *repair* the result. Using this technique, we apply crossover and mutation to literal representations, but before passing the children back to the population, we repair any illegal aspects of the chromosome that have been generated. In the case of the graph coloring problem, illegally colored nodes can be fixed by collecting them into a list, ordering them in some way, and applying the order-based decoding strategy to them. That is, take each one in turn and give it the first legal color it could have, and give it no color if all the colors have been used on nodes it is linked to.

Other schemes for repair strategies exist, but this is a new and relatively undeveloped field. If you are thinking of trying literal encodings with repair mechanisms to fix genetic defects, you will probably be breaking some new ground. For further discussion of these points, see the chapter by Liepins and Potter and the chapter by Whitley, Starkweather, and Shaner in Part II.

THE OPTIMALITY TEST FOR ORDER-BASED REP-RESENTATIONS

The order-based encoding technique has an important advantage over literal encoding techniques in that it rules out a tremendous number of suboptimal solutions to the graph coloring problem. Using a literal encoding, for example, we can generate solutions with nodes that are uncolored, even though these nodes could legally be colored. This cannot happen with the decoder we are using, applied to chromosomes that are permutations of the nodes on the graph.

If you use an order-based representation with a decoder on a problem of your own, you will probably cut down the number of possible solutions that the genetic algorithm will be considering. A good many solutions with low evaluations will be eliminated. For some problems it is possible to eliminate the optimal solution as well. One important question to consider is, given your problem, can its optimal solution be encoded using an order-based representation?

The answer to this question is "yes" for the graph coloring problem. The proof is as follows. Suppose that we have an optimal coloring of the nodes of a graph and suppose that we have ordered the legal colors in some way. We construct a chromosome that has the same evaluation as the optimal coloring as follows.

Consider all the nodes of the graph in order. Give each node that could legally be colored with an earlier color that earlier color. (This won't change the score made by the coloring, since nodes that were already colored are only changing their color. No new colors are being added or deleted.)

Now we build the chromosome that encodes our recolored graph. It is constructed in the following way: First, list all nodes of the recolored graph that have the first color. Second, list all nodes that have the second color. Continue in this way through the last color. Then list all nodes that are uncolored.

The decoder will operate on this list of nodes to give each node exactly the same color it had in the recolored graph. This happens because the nodes have been recolored so that each had the first legal color it could have. Since the decoder gives each node the first legal color it can, the decoder will reproduce colorings of the recolored graph exactly.

For the case of the graph coloring problem, the optimal solution to the problem can be encoded on an order-based chromosome. This is not true for all problems, and you will want to know when it is not. The U S West problem described in the chapter by Cox, Davis, and Qiu, for example, is one for which order-based encodings are not guaranteed to encode optimal solutions in all cases, even though these representations appear to do quite well on the kinds of situations that are encountered in ordinary practice.

SUMMARY

The principal points addressed in this chapter were the following:

- Some greedy optimization algorithms operate on permutations of lists of elements.

- Hybridizing such algorithms with a genetic algorithm leads to the use of order-based representations.

- Order-based representations require new operators such as uniform order-based crossover and scramble sublist mutation.

- Order-based genetic algorithms are quite different from genetic algorithms that use literal representations, and the optimization behavior of these two types of algorithms varies widely.

- Use of an order-based representation may preclude finding the optimal solution to a problem.

REFERENCES

Oliver, I. M., D. J. Smith, and J.R.C. Holland. (1987). A study of permutation crossover operators on the traveling salesman problem. In John J. Grefenstette (ed.), *Genetic Algorithms and Their Applications: Proceedings of the Second International Conference on Genetic Algorithms*. Hillsdale, N. J.: Lawrence Erlbaum Associates.

7

Parameterizing a Genetic Algorithm

The genetic algorithms we have investigated in the preceding chapters of this tutorial have had a variety of parameter settings. In this chapter you will learn how some of those parameter settings were derived. You will also learn some rules of thumb and some automated procedures for deriving parameter settings for your own genetic algorithms.

PARAMETERIZATION TECHNIQUES

Researchers have used four techniques to find good parameter settings for genetic algorithms:

1. *Carrying out hand optimization.* Because parameterization can be a time-intensive task, Ken DeJong devoted a large proportion of the work in writing his dissertation to finding parameter values for the traditional genetic algorithm which were good across a set of numerical function optimization problems. This set of problems, the DeJong test suite, was origi-

nally described in DeJong (1975). DeJong worked out parameter values for single-point crossover and bit mutation by hand. He held the population size constant at 100 individuals.

2. *Using the genetic algorithm.* DeJong's parameter settings were the standards in the field for many years. In 1986 John Grefenstette used a genetic algorithm to evolve good values for these parameters, again setting population size to 100. His parameter settings, described in Grefenstette (1986), outperform DeJong's hand-derived parameter settings on the same test problems, but the improvement is slight.

3. *Carrying out brute force search.* David Schaffer, Lawrence Eshelman, Richard Caruana, and Rajarshi Das sampled the possible parameter settings across a plausible range of values and reported the result in Schaffer et al (1989). They used the DeJong test suite and some additional problems (including f6) for their tests. After consuming more than 12 months of CPU time in their tests, they were able to show that the optimal parameter settings vary from problem to problem, but that robust parameter settings found by their search performed well across the range of problems they considered. Note that these parameter settings also depend on a population size of 100, binary representation, and the one-point crossover and bit mutation operator. Note, too, that the representation technique they used was *Gray coding* rather than the binary encoding technique we have been discussing here.

4. *Adapting parameter settings.* I have experimented with *adapting* the operator fitness parameters throughout the course of genetic algorithm runs. This adaptive technique is described in the next section.

The first three techniques can consume a good deal of time and resources. In fact, the time and resources required to find optimal parameter settings for a problem domain using these techniques are often orders of magnitude greater than the time one plans to spend solving problems in the domain. For this reason these techniques are generally used to find parameter settings that will be robust across a variety of problem types, instead of parameter settings that are optimized for a particular problem domain.

Unfortunately, the robust parameter settings derived for the traditional genetic algorithm by Dejong, Grefenstette, and Schaffer, Eshelman, Caruana, and Das will not be robust when the genetic algorithm is modified in many of the ways we have considered in this tutorial. Thus, in order to build a genetic algorithm that is tailored to a particular domain, we may be faced with the problem of deriving good parameters for the new genetic algorithm. This is another reason why many genetic algorithm researchers prefer to use the robust traditional algorithm rather than adopting their algorithm to each new problem.

We can parameterize our genetic algorithms with any technique mentioned above. Operator weight settings, however, are quite difficult to optimize, especially if we are altering them over the course of the run. For this reason, I have frequently used the fourth technique to find good weight settings, while holding the other parameter values to robust settings.

ADAPTING OPERATOR FITNESS

Given that parameterization of the relatively simple traditional genetic algorithm can be difficult, how can one parameterize more complicated genetic algorithms that use different representations, different operator sets, and different population sizes?

Until 1989, I usually used hand optimization unless there was enough time to use a genetic algorithm to optimize parameter settings on a domain. In 1989 I developed an efficient way to find reasonable operator fitness parameter values. This technique, which I call *adaptive operator fitness*, was described in Davis (1989), and we will review it here.

In an earlier chapter we saw that good values for operator fitness are not flat; rather, they change over the course of a run. There we used a linear interpolation method to track these changes. The optimal values do not change linearly, however; instead, they tend to move in curved trajectories as the run proceeds.

The technique of adaptive operator weighting described in this section alters operator fitness values so that they lie close to these optimal trajectories. It does so by observing, as the run proceeds, how well each operator has been doing recently. Periodically, operator fitnesses are adjusted so that the fitnesses reflect recent operator performance. The precise algorithm for fitness adjustment has three parts. The first part, a technique for measuring operator performance, is described in Figure 7.1.

This procedure allows us to measure the performance of any operator over a recent interval. Put intuitively, an operator's performance is the amount of improvement in the population the operator is responsible for, divided by the number of children the operator created.

The procedure of passing credit from a child back to its parents is necessary in order to reward operators that may set the stage for creating beneficial individuals without making any improvements themselves. The inversion operator mentioned in Chapter 1, for example, will never alter a chromosome's evaluation, although it may improve the performance of

Computing Operator Performance

1. Whenever a child is created, record who its parents were and the operator that created it.

2. Whenever a child is created that is better than the best member of the population, give it an amount of credit equal to the amount that its evaluation exceeds the best evaluation.

3. When a child is given credit for having a better evaluation, add a portion of that credit to the credit amounts of the child's parents, then to their parents, and so on. The portion of that credit to be passed back and the number of generations to pass back are parameters of this algorithm.

4. To compute the performance of an operator over an interval of a certain length, sum the credit of each child the operator produced in that interval and divide by the number of children. The length of the interval is a parameter of this algorithm.

Figure 7.1: The operator performance technique

Adapting Operator Fitness

Assume that operator fitnesses always total 100. Let x be the amount of operator fitness that we wish to adapt. Then:

1. If all operator performance measures are 0, leave the operator fitnesses as they are.

2. Otherwise, multiply the list of operator fitnesses by a constant factor so they total 100 - x. Call this new list of fitnesses the base fitnesses.

3. Form a list of operator performance measures over the recent interval.

4. Multiply the list of operator performance measures by a constant factor so that they total x. Call this list the list of adaptive fitnesses.

5. Sum each operator's base fitness and its adaptive fitness to derive its new fitness.

Figure 7.2: Adapting operator fitness

one-point crossover on the chromosome's descendants. Mutation operators, too, can create children that are different from their parents and have a good feature as well as a bad feature, so that they are worse than their parents. Crossover may be able to splice the bad feature out, creating a child with only the good feature that is better than any other member of the population. It is important that the initial mutation that created this good feature be rewarded when this happens, since rewarding crossover alone will cause crossover to be a dominant operator, and crossover cannot add diversity to a population that has converged.

Once we know how to measure operator performance we can alter operator fitnesses to reflect this performance. The procedure we will use for adapting operator fitnesses in accord with performance is shown in Figure 7.2.

The base fitnesses are the old fitnesses minus the amount to be adjusted. The adaptive fitnesses are fitnesses based entirely on recent operator performance. The sum of the base fitnesses and the adaptive fitnesses yields

---

Initializing Operator Fitness

1. Set all operator fitness values to be identical, unless you know better initial values for them.

2. Initialize the population and store away each population member and its evaluation.

3. Run the genetic algorithm on the initial population and these operator fitness values to create x new population members, where x is the size of the interval between adaptations of the fitness values.

4. Adapt the operator values and store the adapted values away.

5. Repeat steps 3-4.

6. Average the sets of adapted operator values derived in steps 1-5 and set the operator fitnesses to equal this average.

7. Using these new operator fitnesses, repeat steps 2-5.

8. Continue replacing the operator fitnesses with the new operator fitnesses until you see negligible differences between the old fitnesses and the new ones.

Figure 7.3: An algorithm for initializing operator fitness

the new fitnesses.

One potential problem with this algorithm is that an operator that is important only at the end of a run may lose all its fitness at the start of the run. In such a case it may never be tried late in the run. The OOGA implementation of this algorithm has a minimum bound on operator fitness values. The adaptation mechanism replaces the old fitness of each operator with its new fitness or the minimum value, whichever is larger.

The techniques for computing operator performance and for adapting operator fitness will cause the operator fitnesses to change in accord with performance over the course of a run. In order to use these techniques, however, we must have initial operator weights. An algorithm for generating initial operator weights is described in Figure 7.3.

This algorithm proceeds fairly quickly for a parameterization algorithm. It can yield good initial fitness values in less time than a single run of the genetic algorithm. Once these values are found, the adaptive operator weighting technique can proceed without further parameterization, as long as robust settings for the other two parameters are used. The values used here—adaptation of 15% of the operator weight at intervals of 50 individuals—have done well over a variety of problems.

STUDYING OPERATORS WITH ADAPTIVE OPERATOR WEIGHTING

The adaptive operator fitness algorithm can be used to study new operators. Suppose, for example, that you have an operator that you suspect may be effective on your test domain. You can evaluate its performance by running two genetic algorithms. The first will include the new operator and the second will not. In each case, use the operator fitness techniques discussed here, so that the operator fitnesses you use will be well-tuned to the interaction of the operators in your operator set. If the genetic algorithm with the new operator outperforms the one without it, you have good evidence that the new operator is effective. If the new operator consistently loses market share to a similar operator, or consistently has its share reduced to the minimum amount, this is evidence that the operator is ineffective.

A variety of potentially useful operators will occur to you when you are considering a new problem domain. Experiments of this kind are efficient and effective techniques for separating the valuable operators from those that are not.

SUMMARY

The principal points addressed in this chapter were the following:

- Setting parameters for a genetic algorithm can be a very difficult and time-intensive task.

- Optimal parameter settings will vary from problem to problem, but robust settings work well across a variety of problems.

- Robust parameter settings can be derived in four ways for the traditional genetic algorithm with a fixed population size.

- Robust interpolated settings will do better than robust flat settings, but no one has carried out a study that determines what those interpolated settings should be, even for the traditional genetic algorithm.

- When the genetic algorithm is changed—when the encoding technique, the operator set, or the selection procedure is changed—the appropriate parameter settings can be drastically modified.

- For this reason, many genetic algorithm practitioners use only the traditional genetic algorithm.

- An adaptive technique exists for finding operator fitness settings quickly and effectively. It can be used to find good parameter values for nontraditional algorithms created for high performance on specific domains.

REFERENCES

Davis, Lawrence. (1989). Adapting operator probabilities in genetic algorithms. In J. David Schaffer (ed.), *Proceedings of the Third International Conference on Genetic Algorithms*. San Mateo, Calif.: Morgan Kaufmann Publishers.

DeJong, Kenneth A. (1975). An Analysis of the Behavior of a Class of Genetic Adaptive Systems. University of Michigan, Ph.D. Dissertation.

Grefenstette, John J. (1986). Optimization of control parameters for genetic algorithms. *IEEE Transactions on Systems, Man, and Cybernetics*, SMC-16(1).

Schaffer, J. David, Richard A. Caruana, Larry J. Eshelman, and Rajarshi Das (1989). A study of control parameters affecting online performance of genetic algorithms for function optimization. In J. David Schaffer (ed.), *Proceedings of the Third International Conference on Genetic Algorithms*. San Mateo, Calif.: Morgan Kaufmann Publishers.

8
Where to Go From Here?

In writing this tutorial I have left out or given scant treatment to a number of important topics. Now that you have finished the tutorial you will be able to redress this imbalance. By now you understand the various techniques that a genetic algorithm can employ and the way those techniques function together. This understanding and the architectural framework used in this tutorial will help you to integrate other techniques into your understanding of the genetic algorithm. My goal has been to demonstrate the effects of such techniques by showing you what happens when they are swapped in and out of the genetic algorithm.

The tutorial has concentrated on three types of genetic algorithm optimization techniques: the robust, traditional genetic algorithm using bit string representation and generational replacement with elitism; numerical optimization using real number representation; and combinatorial optimization using order-based chromosomes. The great majority of genetic algorithms that have been created for optimization fall into one of these three categories. If you wish to create a genetic algorithm to solve an optimization problem of your own, you can do so simply by making small modifications to the genetic algorithms described in the text and contained on the computer diskette. This is probably the best way to understand how genetic algorithms really solve optimization problems.

You will probably wish to continue your study of the genetic algorithm field. I would like to highlight three areas that are currently the subject of considerable research in the field. If you read further, you will encounter these topics again and again.

- *Classifier systems.* Classifier systems are machine learning systems originated by John Holland that evolve coadapted sets of rules that can function in interesting and complex ways. Classifier systems are a lively area of research, although this area is not as mature as the optimization technology that we have studied in this tutorial.

- *Performance analysis.* The analysis of genetic algorithm performance has been of central concern to genetic algorithm researchers since the field's beginning. I have largely ignored that line of work here, concentrating instead on an empirical presentation of the operation of genetic algorithms. A growing body of research both within and outside the genetic algorithm field is now focusing on the factors that make problems easy or difficult for optimization algorithms to solve. You may wish to study that research as more and more effective tools for analyzing optimization and learning performance are developed.

- *Parallel genetic algorithms.* Genetic algorithms are highly parallel in operation, and many researchers have studied techniques for implementing them on parallel computers. The proceedings of the last two genetic algorithm conferences contain a number of papers on this important topic. This research suggests that even if we are not using a parallel computer for optimization, we may enhance performance if we simulate parallel runs by maintaining multiple populations and controlling the interactions among them carefully. I expect that this research will soon influence the way we think about the standard genetic algorithm.

There is a good deal more to say in the area I have been discussing—that of improving the performance of genetic algorithms on real-world optimization problems. In fact, we have only scratched the surface. But if you have followed the thread of thought in this tutorial, you now know enough to apply genetic algorithms for yourself and to understand the many techniques that have been developed for genetic algorithm use.

In Part II of this book you will be in the capable hands of a variety of genetic algorithm practitioners who will report to you their experiences as they have applied genetic algorithms to real-world problems. As yet, very few genetic algorithms have actually been fielded. As you will see in Part II, however, a variety of genetic algorithm application projects are in progress, several of which will no doubt result in fielded genetic systems in the near future. I hope that you will be a creator of one of the early successes.

A final note. I am worried that by wedding this tutorial to artificial problems I may have divorced the material from its intrinsic excitement. To correct this defect a bit let me note that genetic algorithms, from their beginning in Holland's research, have inspired passion in their adherents. Perhaps this is because they surprise us in interesting ways and because they exemplify processes we believe may have led to our own existence. Whatever the reason, there is something profoundly moving about linking a genetic algorithm to a difficult problem and returning later to find that the algorithm has evolved a solution that is better than the one a human found. With genetic algorithms *we* are not optimizing; we are creating conditions in which optimization occurs, as it may have occurred in the natural world. One feels a kind of resonance at such times that is uncommon and profound. This book has done its job if you come to feel it too.

Part II
Application Case Studies

9

Overview of Part II

Part II of this book contains chapters by authors who describe their applications of genetic algorithms to real-world problems. These chapters contain material covering a wide variety of problem types, approaches to analyzing the problems, and genetic algorithm solutions. Here you will find an overview of the type of problem approached in each chapter, the type of genetic algorithm used to solve it, and other interesting features of the authors' work.

The content of these chapters is fairly wide-ranging. At one stage in the preparation of this book I attempted to group them in a content-associated way. But this collection is quite diverse and it has resisted my attempts to impose order. I am embarrassed to report that you will find the chapters arranged alphabetically by the last name of their first author.

Chapter 10, "Genetic Algorithms in Parametric Design of Aircraft", by Mark F. Bramlette and Eugene E. Bouchard, deals with a design problem. In this chapter the authors discuss optimizing aircraft designs when the task is posed as that of optimizing a list of parameters. The authors have approached the problem with a number of optimization algorithms, including a genetic algorithm using real number representation. They discuss the performance of each algorithm and describe some innovative techniques used in their quite successful genetic algorithm, including the technique of generating a large number of initial population members and then working with only the best ones.

Chapter 11, "Dynamic Anticipatory Routing in Circuit-Switched Telecommunications Networks", by Louis Anthony Cox, Jr., Lawrence Davis, and Yuping Qiu, describes the results that have been derived thus far in a two-year collaboration between Tony Cox's operations research group at U S West and me. The objective of the study is to optimize the routing of telephone messages in a telephone network in order to minimize costs to U S West. We compare the performance of an order-based genetic algorithm like that in Chapter 6 with several other optimization techniques on this problem. In our discussion of the results, we conclude that the genetic algorithm is a highly successful technique when the problem is complex, but hybridization of these algorithms can lead to better performance than using any of them in isolation.

Chapter 12, "A Genetic Algorithm Applied to Robot Trajectory Generation", by Yuval Davidor, shows how to apply genetic algorithm techniques to the task of planning the path which a robot arm is to take in moving from one point to another. Davidor uses variable-length chromosomes in his solution, and devises some novel and interesting crossover operators for his operator set. A revised and updated version of Davidor's dissertation is available in book form.

Chapter 13, "Genetic Algorithms, Nonlinear Dynamical Systems, and Models of International Security", written by Stephanie Forrest and Gottfried Mayer-Kress, concerns a problem posed by current research in chaotic models of real processes. Mayer-Kress has been profiled in the popular press recently because his chaotic models of international arms races and economic competition seem to model some features of the real-world processes better than more traditional models have done. In 1989 and 1990 Stephanie Forrest, then at Los Alamos National Laboratory, used a genetic algorithm to find good settings of the parameters of Mayer-Kress's models in order to enhance their performance on the models. The results are interesting, and may lead to other projects along similar lines.

Chapter 14, "Strategy Acquisition with Genetic Algorithms", by John J. Grefenstette, describes Grefenstette's experiments with SAMUEL, a genetic algorithm that learns techniques for maneuvering a simulated airplane in order to evade a simulated missile. The genetic algorithm he describes employs several techniques of interest, including variable-length chromosomes composed of rules that form a production system. A chromosome is evaluated by using those rules to maneuver the airplane in simulated interactions between airplanes and missiles. Grefenstette has built knowledge of the production rule domain into his operators in clever ways. In the chapter he describes several other techniques that are unique to his work and that appear to be very effective.

Chapter 15, "Genetic Synthesis of Neural Network Architecture", by Steven A. Harp and Tariq Samad, describes techniques for encoding neural network architectures on binary chromosomes. The authors use variable-length chromosomes and a variety of other novel techniques. Recently a number of interactions have taken place between researchers in the neural network and genetic algorithm fields. This paper describes what in my view is one of the most fruitful of those interactions. If you are interested in learning how to combine neural networks and genetic algorithms, this chapter is a very good place to begin.

Chapter 16, "Air-Injected Hydrocyclone Optimization Via Genetic Algorithm", by Charles L. Karr, describes the solution of a design problem by a genetic algorithm using the bit string representation technique. Like Bramlette and Bouchard, Karr represents the design of an air-injected hydrocyclone as a list of parameters. One novel and interesting feature of his approach is the use of a new operator called "simplex reproduction". Karr shows that a genetic algorithm using this operator is quite effective as a search technique for finding design parameter combinations. Karr was a student of David Goldberg, and he seems to be continuing in a tradition Goldberg pioneered—Goldberg's dissertation work in a Civil Engineering Department had to do with controlling the flow of gas in a simulated pipeline with a genetic algorithm. Karr, too, has a talent for finding promising genetic algorithm applications in engineering domains.

Chapter 17, "A Genetic Algorithm Approach to Multiple Fault Diagnosis", by Gunar E. Liepins and W. D. Potter, discusses the use of a genetic algorithm for finding the most plausible combination of causes for alarms in a microwave communication system. The chapter contains an excellent overview of techniques for incorporating domain constraints in a genetic algorithm. The authors use binary chromosomes to represent solutions to a problem that they show is a type of set covering problem. In this chapter they show how to incorporate knowledge about set covering optimization into their genetic algorithm in novel ways, yielding a high-performance hybrid solution to the problem. Gunar Liepins has been an ambassador between the operations research and genetic algorithms worlds for some time, and this chapter reflects his expertise in both domains.

Chapter 18, "A Genetic Algorithm for Conformational Analysis of DNA", by C. B. Lucasius, M. J. J. Blommers, L. M. C. Buydens, and G. Kateman, describes the development of a genetic algorithm for determining the structure of a sample of DNA based on spectrometric data about the sample. The chapter describes an interesting "cascaded" evaluation technique that greatly enhances the efficiency of their evaluation function. The authors use bit strings to encode molecular structures. Their evaluation function measures the degree to which each decoded structure conforms to the data that have been collected about the sample. The genetic algorithm evolves a description of molecular structure that is in agreement with the data collected. The problem of determining biomolecular structure occupies a central position in the worlds of fundamental and applied chemistry today. The authors' research suggests that genetic algorithm techniques can be of value in solving this problem.

Chapter 19, "Automated Parameter Tuning for Sonar Information Processing", by David J. Montana, describes the application of genetic algorithms to two problems associated with interpreting passive sonar data. The first is a parameterization problem. To solve it Montana used a real-number version of OOGA to find good parameter settings for algorithms employed early on in the process of interpreting sonar data. The second problem Montana discusses is a classification problem. For this problem a genetic algorithm was used to train neural networks that classified sonar signals in various ways. In this second system Montana and I experimented with a number of domain-based operators, including the use of back propagation—a neural network technique—as a genetic algorithm operator. This chapter is useful if you are interested in hybrid genetic algorithms, real number representations for parameterization, or neural networks.

Chapter 20, "Interdigitation: A Hybrid Technique for Engineering Design Optimization Using Genetic Algorithms, Expert Systems, and Numerical Optimization", by David J. Powell, Michael M. Skolnick, and Siu Shing Tong, compares a number of optimization techniques on a variety of design problems. The authors show that their genetic algorithm performs well across a range of design problems, although no single algorithm uniformly dominates the others. One of the most interesting features of this chapter is the authors' use of a technique they call interdigitation, which alternates between phases of rule-based optimization and genetic optimization. The authors' account of hybridization by division of labor is fascinating and provocative.

Chapter 21, "Schedule Optimization Using Genetic Algorithms", by Gilbert Syswerda, describes the application of a genetic algorithm to the problem of scheduling activities in a laboratory in which each activity may affect the others in a variety of ways. Syswerda has been implementing this system under contract to the U. S. Navy. The genetic algorithm uses an order-based chromosome like that in Chapter 6 to represent its schedule. The chromosome is decoded with a decoder that incorporates a good deal of knowledge about the scheduling domain. Syswerda's chapter also discusses some of the factors to consider when optimizing in the scheduling domain. This chapter is a very good one to read if you are interested in order-based representations and/or schedule optimization.

Chapter 22, "The Traveling Salesman and Sequence Scheduling: Quality Solutions Using Genetic Edge Recombination", by Darrell Whitley, Timothy Starkweather, and Daniel Shaner, discusses several interesting topics. The authors describe a technique for solving the traveling salesman problem—a well-known combinatorial optimization problem. Their solution includes novel and ingenious representation, crossover, and repair mechanisms. The authors also show how similar techniques can be applied to the problem of scheduling a Hewlett-Packard production line. This chapter and the previous one by Syswerda both apply genetic algorithms to a scheduling problem. Syswerda uses an order-based representation, and Whitley, Shaner, and Starkweather use a more literal representation. I don't believe that anyone at present can say which strategy for solving scheduling problems will be best on which types of problems. We will learn a good deal from work like that reported here, in which the authors advance the state of the art in literal representation technology.

10

Genetic Algorithms in Parametric Design of Aircraft

Mark F. Bramlette and
Eugene E. Bouchard

INTRODUCTION

The first step in constructing methods for aircraft optimization is to consider the general nature of the problem of engineering design (see Bouchard, Kidwell, and Rogan, 1988). Fundamentally, engineering design is the translation of some set of functional desires into a set of instructions that can be used to "construct" an object that satisfies those desires. In practice, the design process typically generates a largely geometrical description, known as a configuration. In this less complete but more common view of design, the configuration represents an implicit set of instructions for constructing the object. The key requirement for a configuration is that it describes an object in terms that allow it to be built without the builder needing to derive information about how it works, how well it works, or what its uses are.

One virtue of this definition of design for the present purpose is that it provides meaningful inputs and outputs for the design process. For example, a complete set of drawings, material lists, and plans is understood

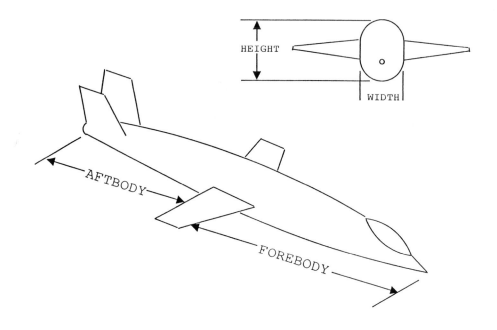

Figure 10.1: Examples of parameters for aircraft design

to be a design for an aircraft. Alternatively, a description of an aircraft in terms of its weight, range, speed, and so on, might specify an aircraft, but it is not a design because the specification is not given in terms that permit aircraft construction.

Parametric design is a subtask of design. The design concept, which is the general type and arrangement of the object being designed (see Figure 10.1) is the starting point of parametric design. In addition, the desires to be satisfied form a set of objective requirements. The design of an aircraft, for example, requires specifying details: Does the aircraft have wings? If so, what are their size, shape, and materials? Does it have engines? If so, what type are they, how big are they, and where are they located? When parametric design starts, many of these decisions have already been made; the results are expressed in the design concept. This concept might specify that the aircraft has wings made of aluminum and a jet engine buried in the fuselage. The object of parametric design is to produce a specific design from the family of designs implied by the design concept. This entails answering questions like: How big should the wing and engine be for a minimum-weight aircraft that meets the range requirement?

A configuration can be specified by a set of symbolic and numerical characteristics that define the objects and relations which comprise the configuration. Selecting characteristics that can be specified independently produces a set of design variables. In parametric design, the design variables define an instance of the design concept being examined. Examples of aircraft design variables are total planform area, tail area, tail location, and engine size. These variables provide the means by which the design can be optimized subject to the design requirements.

Specification of a measure of merit or objective function is central to the parametric design process. It is established by the desires for the object being designed. Only one measure of merit can be optimized at a time (see Keeney and Raiffa, 1976). If a number of characteristics are considered to be important, they may be combined in the objective function using some weighting system. Often, the weightings are not known in advance. They are established from the design characteristics and the designer's judgment. Even though the requirements drive the design, moving away from the optimum frequently produces improvements in other areas which turn out to be too substantial to ignore.

The measure of merit and design requirements together constitute the desires for the object being designed. The requirements can be either quantitative or qualitative, and they can deal with any aspect of the performance, geometry or materials that comprise the design. They often take the form of inequalities. The goal of parametric design is to obtain values for the design variables that satisfy the requirements and optimize the measure of merit.

The range spanned by the objects and relations used to define configurations is known as the design space. Generally, the design space is used as a means of limiting the objects used in the configuration to things that exist or can exist. For example, a list of acceptable materials limits the design space in one direction. The design space and the design requirements often overlap.

Problem Formulation for Combinatorial Optimization

For a design process to be implemented, an appropriate representation of the design problem must be chosen. An important representational issue is the treatment of design requirements or constraints. Constraints can be implemented explicitly, that is, to limit the values of design variables so that only designs that satisfy all constraints are considered. Alternatively, constraints can be implemented through penalty functions that penalize solutions based on the extent that they violate constraints. It is sometimes possible to solve constraint equations or inequalities for some parametric values as functions of other parameters, decreasing the dimensionality of the search space. Although the approaches are ultimately equivalent, some may be more easily implemented or computationally efficient than others.

For example, a concept may involve a tail mounted on the fuselage. The position of the tail is unrestricted as long as it stays on the fuselage and does not interfere with the wing. Hence, the tail position is a constrained design variable. The constraint may be imposed by defining the tail position in terms of the fuselage, and limiting the range of the variable, or it may be imposed as a design requirement. In this example, the first approach is probably easier to implement.

As another example, consider the size of a wing structural component. Several implementations are possible.

- The size of the component can be treated as a design variable, with a design requirement that the component should not break.

- A more abstract design variable can be created which specifies an increment over the minimum required size of a component that allows it to withstand the expected level of abuse.

- The size of the component can be calculated from the weight of the aircraft using relations that size the component so that it will not break.

The first option frees the designer to explore the effects of the component's size. For example, designing a component to be stronger than necessary might have beneficial side-effects, but the additional weight and cost must be considered. The second option still provides design freedom but has the advantage of removing a fairly tedious design requirement at the cost of a more abstract design variable. The third option removes the component from the designer's consideration, at the sacrifice of some design freedom.

Another issue in casting the design problem in a form suitable for combinatorial optimization (CO) is handling continuity in design variables. Usually, it is possible to approximate continuous variables discretely without adversely impacting the quality of the results. For most physical systems, small nonzero perturbations have a negligible impact on the design's utility, an impact less than the errors in our predictions.

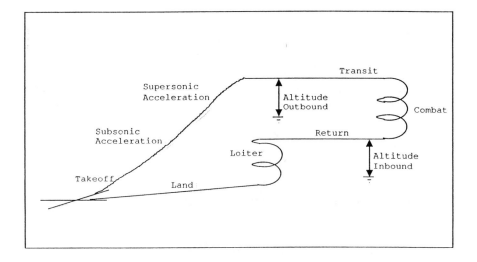

Figure 10.2: Outline of a fighter aircraft mission

Specific Problem Addressed in This Application

The problem addressed here is parametric design using a design concept typical of modern fighter aircraft. This problem is representative of real-world aircraft conceptual design problems. It is similar to practical problems that are currently being solved using the methods described here. The aircraft takes off (see Figure 10.2), accelerates subsonically and then perhaps supersonically, transits to an engagement area, conducts combat operations, returns to base, loiters, and lands. The parameters describing the mission include the altitude, duration and velocity of those activities, range, and acceleration and maneuverability requirements for combat. Representative parameters of the aircraft's geometry and configuration include fuselage width, height, and length; wing planform and thickness characteristics; vertical and horizontal tail surface dimensions; and engine size (that is, aircraft thrust-to-weight ratio). Parameters of the design concept include unscaled engine weight and dimensions, and weights of aircraft subsystems and the weapons complement. The total number of parameters for this problem is fifty.

The associated computations address the geometry of the resulting design, predictions for aerodynamic properties including lift and drag, weight estimates by aircraft component, and analyses of fuel consumption for each element of the mission. Aircraft cost is closely related to weight. Any residual weight in an aircraft at the end of its mission is chiefly unspent fuel. This weight can easily be converted into a feasible reduction in initial weight. Consequently, the object function chosen maximizes residual weight (subject to constraints discussed below), leaving the last easy step in the optimization to the designer. The objective function included penalties to assure that the thrust required for each leg of the mission did not exceed the thrust available on that leg. The penalty is not a fixed amount; it is proportional to excess thrust, if any. This provides information to the search methods about which direction of motion is favorable. Another penalty was applied to reflect the ability of the aircraft to change its energy state, which determines its rates of climbing and accelerating.

Careful formulation of the objective function is critical to achieving worthwhile results. In solving the aircraft design problem just described, the amount of fuel burned while cruising must be calculated. This is done using the requirement for equilibrium in level flight: that is, thrust and drag must be equal at cruise speed. It is implicit in this requirement that the engine is in fact capable of producing the required thrust. In an early formulation of the objective function, this requirement was not explicitly incorporated. As a result, the optimization process discovered that the engine could be made smaller, thereby saving weight, by cruising at more than 100 percent of the engine's rated thrust and "generating" fuel, but that is impossible. The search algorithm was resourceful in solving the problem as stated, but the answer was useless because of this oversight in formulating the objective function.

CUSTOMIZING THE GENETIC ALGORITHM

Our implementation of the genetic algorithm is designed to help determine the fastest search method for our applications through comparing the genetic algorithm with other CO methods. Alternatives (Table 10.1) include simulated annealing, hill climbing, and random search. No method is faster than each of the others for all problems. Any problem addressable by one method is addressable by them all; only their efficiencies vary. Choosing the best method for a particular problem might be based on knowledge of problem topology, but this is rarely known in detail for realistic, high-dimensional problems. Comparing methods by testing them on real problems is more practical. A separate software module was developed for each

Stochastic Hill Climbing
Stochastic Hill Climbing and Hill Climbing
Stochastic Hill Climbing and Next-Ascent Hill Climbing
Stochastic Hill Climbing and Simulated Annealing
Iterated Hill Climbing
Iterated Next-Ascent Hill Climbing
Iterated Genetic Algorithm
Iterated Genetic Algorithm with Stochastic Hill Climbing
Iterated Simulated Annealing
Iterated Simulated Annealing and Hill Climbing
Iterated Simulated Annealing and Next-Ascent Hill Climbing
Random Search
Random Search and Stochastic Hill Climbing
Random Search and Hill Climbing
Random Search and Next-Ascent Hill Climbing
Random Search and Simulated Annealing

Table 10.1: Optimization Methods in the Search Suite

of various search methods and combinations and generalizations of them. These search modules, referred to collectively as the *search suite*, have the same interface with the software that defines particular search problems. Each search method and each problem statement is a separate module; any method can be used with any problem without recoding either. Where the genetic algorithm's usual assumptions are more restrictive than the assumptions of the other methods (requiring binary coding or nonnegative objective functions, for example), we have adapted the genetic algorithm to use the more general assumptions; this facilitates comparing methods.

The genetic algorithm, usually limited to maximization and to nonnegative objective functions, was revised to remove these assumptions. Search-suite users simply specify whether maximization or minimization is required. A scaling parameter was introduced to allow negative values in objective functions; fortuitously, it also combats premature convergence (loss of diversity in the population). The competition for producing the next generation in a genetic algorithm may be accomplished as follows: For each generation, compute a normalized score for each member by dividing its score by the average score of its generation; use the integer part of the normalized score to determine a guaranteed number of copies of the member in the next generation and the fractional part as the chances of putting one more copy in. This procedure depends on the objective function being nonnegative. (A positive average score and a negative unnormalized score produce a negative normalized score; then the procedure cannot be followed.) The scaling factor solves this problem. It is chosen by the user to be greater than zero and less than one; it indicates the part of the variability in normalized scores that is due to the difference in scores. The scaling produces normalized scores so that, for example, if the scaling factor is 0.6, the difference between the greatest and the least normalized scores will be 0.6 times the greatest normalized score. Low scaling factors allow lower scoring members a better chance of reproduction, which helps avoid premature convergence. Given a group of scores and the scaling factor, the procedure is: Compute the average score in the group and an offset given by

((Largest-Score - Smallest-Score) / Scaling-Factor) - Largest-Score

Then, for each score, compute the normalized score as

(Score + Offset) / (Average-Score + Offset)

The integer and fractional parts of normalized scores are used as before to determine a guaranteed number of copies and the chances of one additional copy in the next generation. Following this procedure can lead to one generation having slightly more or fewer members than the parent generation just due to the randomness of producing the additional copies. In our implementation, each generation is forced to have the same size as the original one through user-chosen options: (1) oversized populations are reduced by eliminating lowest scoring members, while undersized populations are increased by duplicating the highest scoring member, or (2) members to copy or to eliminate are chosen randomly.

In our genetic algorithm, each population member is represented as a list of nonnegative integers; each integer gives a design parameter's value, possibly scaled and translated. Since the user may restrict these integers to zero and one, binary coding is possible. Genetic algorithms generally benefit from an operator that makes small rather than gross changes in a policy, in order to refine solutions. Where binary strings are used, gray scale coding has been introduced so that a small change in a policy is more likely to correspond to a small change in the number assembled from the bits. This may make mutation more effective as a refinement operator. We avoid binary coding partly because the problem of mutations producing drastic changes in a parameter value does not appear to be solved adequately by gray scale coding but mainly because binary is not required by any alternative CO method. In our aircraft design applications, no fixed cardinality was imposed on all variables. Each variable is allowed as many values as the user wants according to his judgment of the desirable resolution for that variable. In the problem discussed below, altitude had thirty-one possible values (from 30,000 to 60,000 feet with 1,000-foot resolution, mapped onto the integers from 0 to 30), while wing loading had seventeen possible values (from 30 to 110 with 5 pounds-per-square-inch resolution).

The search suite includes a variant of the genetic algorithm in which the mutation operator is a run of stochastic hill climbing (SHC) (see Ackley, 1987). A member chosen for mutation is used as the starting point of the SHC; the best policy found during that search is put into the next generation as the mutated version. This provides a refinement operator that works better than simpler mutation (replacement by a nearest neighbor) in our aircraft design problems. A biological analog might be a subpopulation that temporarily exploits a favorable environment without competition before rejoining the population.

Both vector and permutation searches are supported in the search suite. Problems in aircraft design exemplify CO problems that require selection of values for some number of parameters. A design is specified as a vector or list of values for each of the parameters, where each value is a number chosen between limits and perhaps subject to constraints involving other parameters. Aircraft manufacturing planning problems are examples of another kind of CO problems; these require finding the optimal permutation of some given set of values. The code for the two kinds of search differs only in the perturbation methods, that is, the way in which a nearest neighbor of a policy is defined and in the implementation of mutation and crossover. A nearest neighbor of a vector policy is defined as any policy produced by adding or subtracting one unit from one of the numbers in the list. For permutation policies, the search-suite user can pick between two definitions of nearest neighbor of a policy according to the problem at hand: a nearest neighbor of a policy is produced either by interchanging the position of two adjacent items or by reversing the order of a subsequence of the policy. In permutation searches, representation is not limited to integers since no arithmetic is done by crossover or mutation. A policy may instead be a list of the names of tasks. The crossover operator for permutation searches must assure that the results of crossover are legitimate permutations, that is, without duplications or omissions. The search-suite method to accomplish this produces offspring in which elements are positioned according to the average of their positions in the two parents.

Objective functions in applications may be undefined at some points in the space being searched, but CO methods generally assume that an objective function is defined for each point in search space. In our applications, however, we found no practical way to recognize combinations of input variables that lead to uncomputability, such as extrapolation beyond table limits. We solved this dilemma by trapping the errors so that unscorable policies could be treated as infeasible.

Search-suite options direct reproduction methods and population size control. In some genetic algorithm implementations, an individual selected for mutation or crossover (a "parent") is put into the next generation, along with its revised descendant, to prevent the destruction of high-scoring individuals. This may accelerate convergence, but it may also promote premature convergence. Our genetic algorithm implementation allows the user to choose whether to preserve parents. This process and the method described above for producing the next generation can result in the population size varying between generations. An issue in genetic algorithm design is whether to use random or deterministic methods to maintain constant population size in each generation. In the deterministic method, oversized populations are trimmed by eliminating low scorers while undersized populations are increased by duplicating high scorers. This may promote convergence. To avoid deterministic decisions, members to eliminate or duplicate may be chosen randomly. Search-suite users may select either approach and determine experimentally which works best on their problems.

Our implementation is an iterated genetic algorithm (see Ackley, 1987) in which each iteration consists of production of an initial population and its subsequent evolution for the user-specified number of generations. Like hill climbing or simulated annealing (see van Laarhoven and Aarts, 1987), any one iteration of the genetic algorithm may produce a population converging to a suboptimal solution. The search-suite user may choose the number of iterations of these methods. This increases the chances of finding optimal solutions.

The method for initializing populations has been generalized. The initial population for a genetic algorithm is usually chosen at random. No known natural event would scramble the values of genes that way. This method of initialization may account for the perception that crossover rather than mutation is the main driver in genetic algorithms. As a possibly better way to initialize populations, our genetic algorithm implementation establishes each member of an initial population as the best policy found in a random search of user-chosen length. Setting the length of the random search at one produces the usual genetic algorithm. Experimentation indicates that a higher value gives better results in some problems. Choosing each member of the initial population as the best of several randomly generated candidates is not the same as seeding the population with some members that are thought to have good scores. Some differences from this seeding in our initialization are that the user does not provide the initial members, premature convergence may be less of a problem, and the expected score of each initial member is the same. In one test problem in which the length of the search was fixed, choosing each initial member as the best found in a random search of length 10 produced a better result than length 2 did. Since the number of function evaluations was the same, using more of them in selecting the initial population forced an end to the search after fewer generations, but it was beneficial even so.

The search suite provides interim and final reports of search progress. The interrupt capability allows users to observe progress in the search, detect a point of diminishing returns, and estimate run completion time. When the search suite is interrupted while searching, it shows an interim report of the best policies found so far in the search and their scores, how many policies have been evaluated, and the search method being used. Optionally, it also shows a search trajectory report that shows each score that is better than any found before it and which trial produced that score. At the start of the search, the user chooses the number of best policies to be displayed. It is independent of the user-chosen population size. The reports produced by an interrupt, updated to their final form, are shown at the end of the search. This includes the best solutions ever found during the search, regardless of whether they are in the final population.

PERFORMANCE OF THE GENETIC ALGORITHM

Assessment is based on statistical methods for comparing alternative search methods. The search suite makes the genetic algorithm, simulated annealing, hill climbing, random search, and variants and combinations of these methods available for problems including optimizing aircraft design. Repetitions of any design problem using any search-suite method generally show variation in results owing to random effects. Statistical methods are necessary to permit sound conclusions on effectiveness in the presence of this statistical noise. We established confidence in assessments of the relative effectiveness of these methods by analyzing experimental results with the t test. In our experiments comparing the genetic algorithm and other methods, we gave each method the same amount of computer time to solve the same design problem. Each experiment using each search method was replicated to provide higher confidence in the significance of observed differences in average final best scores.

Most of the computer time in these searches is used in evaluating objective functions, rather than in code specific to any search method. Fixing how many policies may be scored is therefore a workable substitute for fixing the amount of time that the computer is permitted for search. The computers used in our experimentation varied in speed because of hardware differences and the level of other work required from them. We therefore used a fixed number of policy evaluations in each replication of our experiments.

Search methods were thoroughly tested on one characteristic parametric design problem. The work on other designs and on manufacturing scheduling problems is underway. No one search method or selection of values for search control parameters is optimal for all design problems. Guidance on the selection of methods and control parameters may become available as more experiments of this sort are conducted. Ideally, a program would determine which method and control parameters to use by conducting its own experimentation with a given problem or by consulting dependable heuristics.

Selecting values for the genetic algorithm's control parameters ("tuning") of the genetic algorithm is problematic. The person who uses the search-suite genetic algorithm chooses values for population size, number of random trials made in selecting each initial member, number of generations, mutation and crossover probabilities, whether to put unmodified sources of mutated or crossed over individuals into the next generation, whether generation size is maintained by random choice or according to score, the fraction of each policy's normalized score to attribute to its raw score, and the number of iterations of the genetic algorithm. Selecting values for these parameters is itself a difficult CO problem that has been addressed by the search suite. Although such automatic tuning has performed better than users' guesses of control values in some problems, it was not used extensively because it requires long computer runs. Users' tuning may improve with experience, but acquiring extensive experience and generalizing from it correctly are also difficult.

RESULTS AND CONCLUSIONS

Eight parameters (Table 10.2) were chosen from the fifty defining the parametric design of a fighter aircraft. These were permitted to vary in the general vicinity of the values suggested by the preliminary manual analysis. The resulting search space contains over 136 billion points.

Parameter Name	Lower Limit	Upper Limit	Resolution	Units
Wing Loading	30.0	110.0	5.0	lbs / sq in
Thrust-to-weight Ratio	1.0	2.0	0.1	—
Length of Aft Body	2.0	10.0	0.25	feet
Length of Fore Body	15.0	55.0	1.0	feet
Fuselage Height	2.0	6.0	0.25	feet
Fuselage Width	4.0	12.0	0.25	feet
Altitude Outbound	30,000.0	60,000.0	1000.0	feet
Altitude Inbound	30,000.0	60,000.0	1000.0	feet

Table 10.2: Free Parameters in the Search

Search Method	Number of Tests	Number of Scorings per Test	Average Score in Pounds	Standard Deviation of Scores
Random	24	4,680	3,553	442
Stochastic hill climbing	24	4,680	4,776	65
Iterated simulated annealing	24	4,680	4,822	0
Iterated genetic algorithm	32	4,680	4,819	5
Genetic algorithm with SHC	32	4,680	4,822	0

Table 10.3: Results of Tests of Search Methods

The stochastic hill climbing method (like simulated annealing, but with a constant temperature) was tested using 10 as the (dimensionless) temperature. The iterated simulated annealing method was tested using a starting temperature of 100 and an ending temperature of 0.1, so it made sixty-five iterations of stochastic hill climbing. Each of these was of length eight, producing 520 function evaluations. This simulated annealing method was iterated nine times to produce 4,680 function evaluations per application (see Table 10.3).

The iterated genetic algorithm was tested using a population size of fifty-two. The variable controlling the number of random trials made to select each member of the initial generation was set to one as in the usual genetic algorithm methods. Each iteration of the genetic algorithm was run from a new, randomly chosen population for 440 generations. Each mutation requires one function evaluation, and each crossover requires two. Mutation and crossover probabilities were set to produce an expected number of function evaluations of 0.1 per population member per generation (two-thirds of them for crossover), or 5.2 evaluations per generation for 440 generations, or 2,288 evaluations over all generations of one iteration. Adding in the fifty-two evaluations used to find the initial population produces 2,340 altogether per iteration. Two iterations produced an expected number of 4,680 function evaluations per application of the genetic algorithm. The dynamic range was set at 0.99 since preliminary exploration indicated that premature convergence was not a problem in this case. Parents were not put into the next generation and population size was maintained by random selection.

In the conclusions that follow, "t" represents the t test statistic for determining whether two populations have equal means (Crow et al., 1960) and "p" is the probability that the observed value of t occurred merely by chance rather than because there is a difference in the means. Thus a very low probability indicates high confidence that the observed difference in means is significant. The random method is not as good for the test problem as stochastic hill climbing ($t = 12.99$, p < 0.0005), iterated simulated annealing ($t = 14.06$, p < 0.0005), or iterated genetic algorithm ($t = 14.03$, p < 0.0005). Stochastic hill climbing is not as good as simulated annealing ($t = 3.42$, p < 0.005) or iterated genetic algorithm ($t = 3.20$, p < 0.005). The iterated genetic algorithm was not as good as the iterated simulated annealing ($t = 2.66$, p < 0.005). The difference in their means (4,822 and 4,819), is so small that its statistical significance is somewhat surprising. This difference probably has no practical significance. However, the difference in standard deviation, which favored simulated annealing, *is* of some practical significance since users prefer predictability in results.

To clarify the relative performance of the genetic algorithm and simulated annealing, one more optimization method was tested. That method is the iterated genetic method with stochastic hill climbing (SHC) as the mutation operator, applied thirty-two times. The parameters controlling this were the same as those used in the iterated genetic algorithm tests except that the number of generations was reduced to 220. This preserved the number of function evaluations at 4,680 while accommodating a mutation that consisted of a run of SHC of length four (requiring four function evaluations) at a temperature of 5. This method produced scores with the same mean and variance as the iterated simulated annealing. For this problem, simulated annealing and the genetic algorithm with SHC produced the best answers of the methods considered. A summary of the results is given in Table 10.3.

REFERENCES

Ackley, D. H. (1987). *A Connectionist Machine for Genetic Hillclimbing.* Boston: Kluwer Academic Publishers.

Bouchard, E. E., G. H. Kidwell, and J. E. Rogan (1988). The application of artificial intelligence technology to aeronautical system design. AIAA-88-4426, AIAA/AHS/ASEE Aircraft Design Systems and Operations Meeting, September 7 to 9, 1988, Atlanta.

Keeney, R. L. and H. Raiffa (1976). *Decisions with Multiple Objectives: Preferences and Value Tradeoffs.* New York: John Wiley and Sons.

van Laarhoven, P. J. M., and E. H. L. Aarts (1987). *Simulated Annealing: Theory and Applications.* Dordrecht: Kluwer Academic Publishers Group.

11

Dynamic Anticipatory Routing in Circuit-Switched Telecommunications Networks

Louis Anthony Cox, Jr., Lawrence Davis, and Yuping Qiu

ROUTING NETWORK TRAFFIC

The network control problem addressed in this chapter is a *combinatorial system control problem* for which conventional approaches such as optimal feedback control design, rule-based expert systems, and real-time distributed agent control have not been very effective. In these problems, short-run control can be improved only by concentrating computational resources on a "hard" local optimization problem, temporarily ignoring changes in the overall problem. But ignoring these changes may degrade long-run performance. In such systems, time spent in optimization is time taken away from monitoring the system. This trade-off between problem-solving and monitoring intrinsically limits the effectiveness of the controller.

As an example of the type of network control problem addressed here, consider a scheduler for a telecommunications network. Call requests arrive at random at the nodes of the network. Each call request is specified by six attributes: its source node, destination node, requested start time, requested duration, bandwidth requirement, and priority class. (A more complete description of the problem is given below.) Let $x(t)$ denote the set of call requests requiring scheduling at time t. Let $U[x(t)]$ denote the set of *feasible schedules* (or *potential solutions* to the scheduling problem) for these call requests. A *schedule* assigns to each request in $x(t)$ a path (a sequentially connected set of links) through the network from the request's source node to its destination node over its requested time interval. Call requests can be assigned to a null path to indicate they are not scheduled to be placed. A path must be maintained for the duration of a call: in circuit-switched networks, as opposed to packet-switched networks, the path used during a call cannot be changed while the call is in progress. A schedule is *feasible* for $x(t)$ if the sum of the bandwidths of all requests assigned to a given link (i.e., assigned to paths that pass through that link) is never greater than the link's capacity (the maximum bandwidth that it can carry at any moment.)

Given any feasible schedule u in $U[x(t)]$, a *net cost* for that schedule can easily be computed as the sum of (1) the net costs (if any) of carrying successfully scheduled calls over their assigned paths; and (2) the penalties associated with call requests that are not scheduled, that is, that are *blocked*. Let $C[x(t), u]$ denote the net cost associated with feasible schedule u for the call requests in $x(t)$. The scheduler's short-run optimization problem is to choose a schedule from $U[x(t)]$ that will minimize $C[x(t), u]$. However, this is in general a very hard (NP-complete) problem. The problem's difficulty is compounded by the timevarying nature of call requests, as discussed below. Suppose that the scheduler spends an amount of time T scheduling the current requests $x(t)$, and let $u[T; x(t)]$ denote the best solution that it has found after T periods. Now $C[x(t), u[T; x(t)]]$ is presumably a decreasing function of T: by spending more time working on problem $x(t)$, the scheduler can find a better (lower cost) solution to it. However, *during the T periods spent in computation, the problem will have evolved from $x(t)$ to $x(t + T)$*, reflecting both the arrival of new call requests over the interval from t to $t + T$ and the fact that call requests with start dates in this interval will already have been missed. Moreover, the assignment of a specific call to a specific path that is "optimal" in $x(t)$ may be nonoptimal in $x(t + T)$. Thus, effort invested in solving $x(t)$ may not reduce the amount of effort needed to solve $x(t + T)$.

These characteristics can arise in a range of other problems, including real-time data fusion and tactical planning for certain C^3I applications; on-line diagnosis and repair of multi-component systems with soft or transient failures; and robot planning to achieve a goal at least cost in a changing environment. Abstractly, consider a controlled stochastic system evolving in discrete time according to the transition function $F[x(t+1)|x(t), u(t)]$, where $x(t)$ denotes the state of the system at time t, $u(t)$ is the input to the system (e.g., a control signal) at time t, and $F[x(t+1)|x(t), u(t)]$ is the probability that the system makes a transition to state $x(t+1)$ starting from state $x(t)$ when the input is $u(t)$. Let $C[x(t), u(t)]$ be the cost per period when the state is $x(t)$ and the input is $u(t)$, and let $U(x)$ denote the set of feasible inputs for state x. Let $u * (x)$ denote the input in $U(x)$ that minimizes $J(x, u) = C(x, u) + V(x, u)$ over all u in $U(x)$, where $V(x, u)$ is the expected discounted (or averaged) sum of all future per-period costs starting from (x, u) and assuming optimal future decision-making. $J(x, u)$ is the *objective function* of the control problem. (This is the standard dynamic-programming formulation of the optimal system control problem in discrete time.)

Finding $u*(x)$ is an optimal control problem. It is a *control problem with time pressure* if, informally, the problem of computing $u * (x)$ takes long enough to solve to optimality (e.g., because it is NP-complete) so that the system is very likely to make a transition to a new state, say x', for which $u * (x)$ is nonoptimal, before $u * (x)$ has been computed. By spending more time thinking about the control problem starting from state x, the choice of u to minimize $J(x, u)$ can be improved—but its relevance to the (evolving) real current state diminishes. This dilemma also arises for systems evolving in *continuous* time, where the state drifts away from its initial value while the controller is considering how to respond. The trade-off between solving the initial optimization problem better and applying it soon enough to be useful is an important feature of real-time decision-making familiar to most human decision-makers.

In slightly greater detail, let A be an iterative optimization algorithm that takes a problem state x and an objective function $J(x, u)$ as input and that produces a sequence $\{u_1, u_2, \ldots\}$ of increasingly good approximations to $u * (x)$ as output. Assume that $J[x, u_{i+1}(x)] < J[x, u_i(x)]$ for each i in this sequence (except the last one, if the sequence converges to $u * (x)$ in a finite number of steps). In other words, A can be viewed as a routine that produces a sequence of *improvements* in the choice of control for x, corresponding to reductions in the objective function starting from x. While A is running, the state of the system may undergo one or more random transitions. The control problem with time pressure using algorithm A amounts to deciding how long to run A on problem x before selecting the best answer discovered so far, applying it to the system, and restarting A on the new state of the system.

The remainder of this chapter focuses on how to develop an "intelligent" controller for making control decisions under time pressure. The call request scheduling problem is used throughout as a concrete application. However, the principles to be developed apply to more general control problems with time pressure.

THE CALL REQUEST ROUTING AND SCHEDULING PROBLEM

The call routing and scheduling problem described in this section is motivated by opportunities to offer new services in the telecommunications industry by offering private network owners the ability to *reserve capacity* to carry certain kinds of traffic (e.g., video conference calls, periodic transfers of corporate data, and so forth) that can be anticipated and planned for in advance. To achieve these benefits, algorithms for *anticipatory* routing and scheduling of call requests are required that take into account available information about future bandwidth requirements. This section formulates the general anticipatory routing problem for circuit-switched networks, stripped of specific applications details.

A telecommunications network contains several Customer Access Points (CAPs) (its nodes) which provide points of entry for customer call requests into the system. A *call request* consists of at least five attributes: a source-destination pair specifying from which CAP and to which CAP the call is to be placed; a bandwidth (bit-rate) requirement for the call; a desired start time; a duration; and (optionally) a priority class or code. Additional call attributes (e.g., required security level, information about the customer, and so forth) can be added to this set without greatly changing the problem. At each CAP, call requests arrive in an apparently random (nonhomogeneous compound Poisson) stream with a time-varying average intensity or rate that reflects underlying, time-varying demand. Given the pattern of call requests arriving at different nodes (CAPs), the routing problem addressed here is to *assign feasible paths to calls in a way that will minimize the (priority-weighted) average number of call requests blocked or denied per unit time*. A call request is blocked if there is no way to place it. This occurs if, on each path connecting the source CAP to the destination CAP, there is at least one link whose capacity will be exceeded during at least one point in the time interval required by the requested call if the requested call's bandwidth is added to the traffic already committed to use that link. Each blocked call generates a certain penalty or *loss*. The routing problem is to dynamically assign paths to calls over time so as to minimize the average loss generated per unit time. More formally, the problem may be described as follows.

Given:

1. A *capacitated network* (N, E, c), where N is a set of nodes, E is a set of bidirectional links, and $c : E \Rightarrow C$ is a capacity assignment function assigning to each link e in E a capacity in C. C is a small, finite set of possible capacity values (e.g., $C = \{1, 24, 672\}$ where capacities are measured in units of equivalent voice-grade lines). A link is specified by the pair of nodes that it joins and by its ID number: the same pair of nodes can be joined by more than one link.

2. A *traffic pattern* $\{\lambda_i(t)\}$ where i indexes *call types* and $\lambda_i(t)$ is the average arrival rate for calls of type i at time t. Arrivals are assumed to occur at each node according to a nonhomogeneous (time-varying) Poisson arrival process. A *call type* is defined by its source node, destination node, size (= required capacity or bandwidth), requested duration, and priority class. (Other attributes may be added.) A *call* is an instance of a call type that has a starting time attached to it.

3. A *call table*, M. Rows of M represent individual calls that have been requested. Columns of M represent call attributes, (i.e., source node, destination node, size, requested start time, requested end time, priority class). Newly requested calls are generated according to the traffic pattern and are inserted into the call table. Calls in progress are deleted from the call table as they terminate. (Note: New requests may be denied unless they have start times at least an hour ahead of the current time.)

4. A *loss function* associating a loss, or penalty, with each call type. Let $w(i)$ denote the loss for call type i. Then $w(i)$ is interpreted as the penalty for each call of this type that is requested but not successfully placed (i.e., that is *blocked*).

Find:

- A dynamic routing policy that minimizes the average loss per unit time from blocked calls.

The terminology is as follows. A *dynamic routing policy* has two parts: (1) a *commitment decision rule* that determines whether to *accept* (i.e., to make a commitment to place) each requested call at the time it is requested; and (2) A *path assignment rule* that assigns a feasible path to each accepted call request before its requested start time. Initial path assignments are tentative and may be revised several times before the call is placed. On the other hand, the system is not allowed to accept a call request without identifying a feasible path that can carry it. Each accepted call is placed along the last path assigned to it before its start date. Thus, these two decision rules take the capacitated network data, the current traffic pattern, the current call table, and the loss function as inputs and produce accept/reject decisions and feasible assignments of paths to calls as outputs. The *performance* of a dynamic routing policy is evaluated by the expected loss per unit time that it generates.

OUTLINE OF THE DYNAMIC ANTICIPATORY ROUTING ALGORITHM

We have developed an approach called Dynamic Anticipatory Routing (DAR) for solving the call request routing and scheduling problem in real time. It makes essential use of the following three concepts: (1) a *statistical* approach can be used to make rapid initial assignments of incoming call requests to paths based only on aggregate statistical information about the current traffic pattern; (2) initial assignments can be *sequentially improved* by searching for the solution to a hard but static combinatorial optimization problem; and (3) the decision to accept the best solution discovered so far and to move on can be guided by *statistical analysis of the sequential improvement process* generated by a given optimization algorithm. These ideas are developed in the following sections. A summary outline of the DAR algorithm follows.

Outline of DAR heuristic for real-time call scheduling

Step 0. (Initialize master strategy.) Compute a master strategy as follows: (a) For each call type, find the few (e.g., 4) shortest feasible paths connecting the source and destination nodes for that call type. (b) A *master strategy* for path-assignment is a static set of path-selection probabilities for call types mapping call type i into feasible path j with probability $p(i, j)$. (If j is busy, $p(i, j)$ is set equal to zero and the remaining path-assignment probabilities are renormalized to sum to 1.) Each possible strategy $[p(i,j)]$ generates steady-state blocking probabilities (via a generalized Erlang fixed-point problem with unequal call bandwidths) and hence an expected loss or penalty per unit time. Step 0 solves a nonlinear program to find the $[p(i,j)]$ pattern giving the least expected loss per unit time within this class of randomized assignment strategies.

Step 1. (Test for changed traffic pattern.)   At the beginning of a cycle, test whether the traffic pattern has changed significantly since the last time a master strategy was computed. If it has, recompute a master strategy for the new traffic pattern (Step 0). Otherwise, continue to Step 2.

Step 2. (Feasibility test for incoming call requests.)   When a new call request r is received, attempt to assign it to a feasible path using the current master strategy $[p * (i,j)]$. [Path j is *feasible* for call request r made at time t for the future interval $(t1, t2)$ if each link in j is currently (at time t) scheduled to have enough residual capacity on it over the interval $(t1, t2)$ to carry r). If a feasible path is available, then *accept* call request r, that is, make a commitment to place it at its requested start time and assign it a path, say $j(r)$, using the current master strategy $[p * (i,j)]$. Append the pair $[r, j(r)]$ to a *pending call list*, denoted by $L(t)$, and return to Step 1. Otherwise, if no feasible path for r can be found, go to Step 3.

Step 3. (Bandwidth-packing.)   Let $L(t)$ be the current pending call list, and let r be a call request that has not yet been accepted or rejected. This step seeks a new set of assignments of paths to call requests in $L(t)$ that will make it possible to also accept call request r. To this end, each current path-assignment pair $[r, j(r)]$ for a call not yet in progress is allowed to be modified to a new assignment, say $[r, j * (r)]$. (As soon as a requested call is in progress, it is removed from the list of pending calls.) The search for a new set of assignments that will allow r to be accommodated is accomplished using a genetic algorithm, as follows:

3.0 (Permutation evaluation.)   Let L denote an arbitrary initial permutation of the call requests in $L(t)$. The loss associated with L is defined as follows. Take the first call request in L and assign to it the shortest feasible path in the pre-screened path set for calls of that type. Subtract its bandwidth from each link in that path. Repeat the process with the next call in L. Continue until all calls have been assigned or until no feasible paths exist for the remaining call requests. The loss associated with L is the sum of the losses from all call requests not assigned under that permutation. The idea of the genetic algorithm is to start with a *population* of alternative permutations and successively to select and combine the best (loss-minimizing) ones to evolve a better permutation.

3.1 (Permutation evolution.)   Generate a random initial population of alternative permutations. Evaluate each one (Step 3.0). Select and carry out genetic optimization using the genetic algorithm described below. Continue to generate, evaluate, select, and combine permutations until a termination condition is encountered (e.g., no improvements found after x CPU-seconds). The best permutation discovered so far is the output of the genetic algorithm and is used to seed the remaining steps.

3.2 (Local permutation improvement.)   Examine all pairwise interchanges of call requests in the incumbent permutation to determine whether a lower loss "adjacent" permutation can be found. If an improvement is found, use it as the seed for another round of this local improvement ("2-opt") step. Continue until a termination condition is satisfied (e.g., no improvements found in y CPU-seconds.)

3.3 (Termination.)   As soon as a feasible permutation is found for the previously accepted call requests plus call request r, send an "accept" message (a confirmation) to the requestor. If no feasible permutation is discovered before termination, send a "reject" (denial) message. Return to Step 1.

The DAR algorithm consists of two main parts, as follows.

- Part 1, called the *master scheduling algorithm*, takes the current pattern of traffic arrival rates, $\{\lambda_i(t)\}$, as input and produces a *master strategy* for routing calls as output. A master strategy is a probabilistic path-assignment rule that assigns a path to each call coming into a node (CAP). Assignments are made using restricted randomization to minimize the expected loss generated by blocked calls per unit time without taking into account the detailed state of the network (e.g., which calls are currently assigned to which links). This procedure involves solving a nonlinear optimization problem similar to those treated by Kelly (1988) but with variable bandwidths.

- Part 2, called the *bandwidth-packing algorithm*, is used to improve the initial path assignments made by the master scheduling algorithm. It takes as inputs

  1. The current state of the network (how much bandwidth will become free on each link, and when, as currently in-progress calls terminate).

  2. A detailed list of call requests that have been accepted so far but that are not yet in progress, as determined from the call table $M(t)$.

3. The current statistical traffic pattern $\{\lambda_i(t)\}$.

From these inputs, the bandwidth-packing algorithm computes detailed path-assignments of network paths to requested calls in an effort to minimize an objective function. A simple objective function for the dynamic problem is the blocking probability for calls generated by traffic pattern $\{\lambda_i(t)\}$, given the currently scheduled commitments of paths—or, equivalently, the currently scheduled availability of bandwidth on various paths over time. A more general objective function is the expected loss per unit time, that is, the priority-weighted call blocking probability, generated by $\{\lambda_i(t)\}$ and by the currently scheduled path commitments (including calls already in progress.)

The master scheduling and bandwidth packing routines are embedded in a *real-time scheduler* that makes accept/reject decisions for individual call requests and that identifies initial path assignments for accepted calls. In the remainder of this chapter, we focus on the bandwidth-packing algorithm, which is the most time-consuming part of the DAR framework.

A Heuristic Approach to Tactical Planning under Time Pressure

The optimal master strategy minimizes expected loss per unit time in a statistical, or expected value, sense for the input data (N, E, c) and $\{\lambda_i(t)\}$. However, it ignores the actual state of the network at each moment, relying instead only on the general network description (N, E, c). By contrast, the bandwidth-packing algorithm uses detailed information about the current and future states of the network to find new path-assignments that will further reduce the expected loss per unit time. This section describes the bandwidth-packing process for a fixed list of calls to be placed. This process is called repeatedly, and the list of calls to be placed is updated periodically in real-time applications, but we will ignore these complications here to better show the essential structure of the algorithm. The bandwidth-packing problem for a fixed list of calls is of great interest in itself across a wide variety of domains where similar techniques can be used.

Let x denote the list of calls to be placed at any moment, each with an associated bandwidth requirement, and let $u(t)$ denote a schedule at time t assigning the calls in x to paths through the network. At time t, x cannot be changed (accepted call requests are locked in), but u can be. As another piece of notation, let $U(x)$ denote the set of feasible schedules starting from state x.

Ideally, an evaluation function $J(x, u)$ for a schedule u starting from a state x might consist of the minimum expected value of the discounted sum of future per-period losses from blocked calls, starting from (x, u) and assuming optimal scheduling henceforth. In practice, this exact evaluation function is much too hard to compute. It is therefore necessary to use a *heuristic evaluation function*, $h(x, u)$, instead. For our purposes, we will let $h(x, u)$ denote the *total profit* from placing the calls in x according to schedule u. This is calculated by assigning a profit of $-w(k) - c(k)$ to each call k in x that is successfully placed, where $w(k)$ is the penalty for *not* placing call k and $c(k)$ is the cost of placing it along the path assigned to carry it (the sum of the link carrying costs over all links in the path). Zero profit is assigned to calls that are not placed; thus, the penalties from these calls are opportunity costs. Let $h(x, u)$ denote the negative of the total profit from applying assignment schedule u to problem x.

The tactical planning problem for improving on the path assignments made by the master strategy can now be described as the following optimization problem: choose a schedule from $U(x)$ to minimize the heuristic evaluation function $h(x, u)$. In general, this is a combinatorially hard problem.

SOLVING THE TACTICAL PLANNING PROBLEM: HEURISTIC RANDOM SEARCH

In principle, the solution to the tactical planning problem starting from a problem state x is a schedule, denoted $u*$, that minimizes $h(x, u)$ over all u in $U(x)$. That is, $u* = argmin[h(x, u) : u$ in $U(x)]$. In practice, $u*$ may be difficult to find. Two heuristics are therefore used to construct a sequence of increasingly useful approximations to $u*$. First, the set $U(x)$ of feasible schedules is *restricted* to a subset of schedules that can be expressed in a particularly simple form. Let $\underline{U}(x)$ denote this restricted subset of $U(x)$. Second, an *adaptive sequential search* of the schedules in $\underline{U}(x)$ is conducted in an effort to find the best one. Generically, let A be an algorithm used to search $\underline{U}(x)$. A can be viewed abstractly as a (possibly random) mapping from $\underline{U}(x)$ into $U(x)$ that takes a finite (but not necessarily deterministic) amount of time to execute. Each execution results in an *improvement* of the current schedule for x. (If this is not true for some sequential search algorithm A', then A' can easily be modified so that only improvements are posted as outputs, thus bringing it into the form discussed here.) From any starting point u_0, repeated applications of the improvement algorithm

A will generate a sequence $\{u_1, u_2, u_3, \ldots\}$ of improved schedules, corresponding to smaller and smaller values of $h(x, u)$. If A is deterministic and $U(x)$ is finite, then the sequence of improvements that it generates via the iteration $u_{i+1} = A(u_i)$ starting from u_0 must eventually converge. Let $A * (u_0)$ denote the best schedule reached by successive improvements starting from u_0. In general, $A * (u_0)$ may be (and for many improvement algorithms and scheduling problems will be) suboptimal in $\underline{U}(x)$.

To avoid becoming trapped in a local optimum, a variety of *random search* algorithms have been developed. These methods include *pure random search* (random sampling and evaluation of points in $U(x)$, assuming that these are easy to generate and evaluate); *simulated annealing; tabu search*; and *genetic algorithms*. The last three all use heuristics to guide the random search process. The last two, tabu search and genetic algorithms, use information collected during the search itself to guide the hunt for improvements; hence, they are *adaptive random search* heuristics. The genetic algorithm approach searches many different parts of the search space $\underline{U}(x)$ simultaneously and thus is particularly well suited for problems with many local optima and little spatial structure to guide the search process toward a global optimum. (This also makes the genetic algorithm suitable for parallel implementation.)

A GENETIC ALGORITHM FOR BANDWIDTH PACKING

Computational experiments conducted at U S WEST Advanced Technologies and BBN Laboratories in 1989 and 1990 revealed that a suitably tailored genetic algorithm could make a highly effective choice for the bandwidth-packing component of the DAR algorithm. Moreover, statistical analysis of the improvement sequences generated by various algorithms provides valuable heuristic information for setting the time budget T allocated to bandwidth packing in real-time applications and for choosing among algorithms. This section explains the genetic algorithm approach to bandwidth packing.

Defining the Search Space $\underline{U}(x)$: Permutation-Based Schedules

Any bandwidth-packing heuristic algorithm, A, has three requirements: (1) a restricted search space, $\underline{U}(x)$, of possible schedules; (2) a method for generating points in $\underline{U}(x)$ to evaluate; and (3) a means of evaluating any proposed schedule, u, in $\underline{U}(x)$. This evaluation mechanism is represented symbolically by the heuristic evaluation function $h(x, u)$. Its construction for the profit-maximization problem has just been described. This sub-section describes a search space $\underline{U}(x)$ and various algorithms for searching it.

For all the bandwidth-packing heuristics considered in this chapter, $\underline{U}(x)$ is constructed as follows. Let L be a list of call requests to be scheduled; that is, L is a permutation of the calls in x. Then the search space $\underline{U}(x)$ corresponding to L can be identified with *the set of all permutations of the elements of list L*. Any permutation of L, say, L', can be interpreted as another schedule by the following "decoding" procedure:

Permutation "Decoding" Algorithm for Building Schedules

1. Take the first call on list L' and assign to it its shortest (or least-cost) path, ignoring all other calls on the list. Subtract its bandwidth requirement from the available capacity of each link along its path.

2. Take the next call on list L' and assign to it its shortest (or least-cost) path in the updated network from which the bandwidth requirements of previous elements on the list have already been subtracted.

3. Repeat step 2 until all calls on L' have been scheduled (or until none of the remaining calls can be placed).

The decoding procedure associates a schedule with every permutation of L. We will refer to these as *permutation schedules*. If there are N calls in L, then the size of the search space $\underline{U}(x)$ is $N!$ Since in many of our computational experiments N ranged from 20 to over 50, the corresponding search spaces for potential schedules (encoded as permutations) are enormous. Two questions arise:

1. How can the set $\underline{U}(x)$ of permutation schedules for a given problem be searched efficiently?

2. Is the best permutation schedule a "good" schedule? If $h* = h(x, u*)$ is the optimal value of the heuristic evaluation function achievable for any schedule in $U(x)$ and $\underline{h}*$ is its optimal value for any schedule in $\underline{U}(x)$, then is the difference $h*-\underline{h}*$ close to zero?

Informally, the first question asks how to find the best permutation schedule, and the second asks whether it is worth finding.

The answer to the second question is that, *although the best permutation schedule can be arbitrarily poor in very small problems, it appears empirically to be optimal or near-optimal in large problems.* The basis for this statement is a set of small test problems run to optimality on networks with 10 nodes or 30 nodes, for which exact solutions can be obtained by integer programming (using the commercial package LINDO). In every case, the best of the bandwidth-packing heuristics that were being compared (described below) found solutions with profits that were identical to or within 1 percent of the optimal profit produced by the exact LINDO solution. Moreover, the heuristics were much faster. For test problems with 50 nodes, integer programming was no longer practical. For large problems, exact solutions could not be found in practical amounts of time. We have therefore not been able to verify the conjecture that $\underline{h} * / h*$ approaches 1 with probability 1 as problem "size" (defined in some appropriate way, e.g., as the average number of pivots needed to reach an exact solution from a random permutation schedule starting point) increases. However, all our experimental evidence so far supports the hypothesis that the best permutation schedule is close to optimal in practical problems.

Heuristics for Searching $\underline{U}(x)$

The final piece needed for the bandwidth-packing algorithm is a method of sequentially searching the set $\underline{U}(x)$. To date, we have studied four heuristic strategies, any of which could in principle be used alone to produce a bandwidth-packing procedure. However, as discussed below, it turns out that a hybrid heuristic is more effective than any single one. The pure heuristics evaluated to date are as follows:

1. *Greedy heuristic.* This algorithm simply rank-orders the call requests in L in order of descending profit. (Other criteria based only on the attributes of each call request, for example, profit to bandwidth requirement could also be used.) The resulting permutation represents the schedule generated by the greedy heuristic. This method is used to seed other methods that sequentially generate improvements.

2. *Opportunistic pairwise exchange ("2-opt").* This is a form of hill climbing in the space of permutations, $\underline{U}(x)$. Given an initial permutation generated by some other method (e.g., by the greedy heuristic or randomly), the 2-opt procedure generates up to $N(N-1)/2$ "adjacent" permutations by swapping the elements in each pair of positions

and evaluating the resulting permutation. One possibility would be to evaluate all possible pairwise swaps and then select the one giving the greatest improvement as the seed for the next round. However, we have found that run time to achieve the best solution ever found can be reduced (by roughly a factor of 3), and the quality of the best solution discovered can be improved slightly by using an *opportunistic* version of the 2-opt procedure. In this version, as soon as an improvement is found, it is immediately used to seed a new round of 2-opt exchanges. The procedure terminates when it completes a round (all pairwise exchanges are made and evaluated) without discovering an improvement.

3. *Uniform random search.* Given a limited budget of CPU time to invest, a direct approach is simply to generate and evaluate as many random permutations as possible in the time allowed, keeping the best one found. This strategy amounts to uniformly sampling the space $\underline{U}(x)$.

4. *Genetic algorithm.* The genetic algorithm that we ended up using in DAR differs from more conventional genetic algorithms in that it is order-based and uses a uniform order-based crossover operator. Like other genetic algorithms, it combines the idea of randomly sampling the search space with the idea of using information from the best solutions found so far to focus the sampling process on the most promising parts of the search space. This genetic algorithm is very similar to that used in the graph coloring chapter of the tutorial in Part I of this book. The principal difference is that an exponential fitness function, rather than a linear fitness function, was used. This technique is described below.

There are two main ideas:

(a) Permutations in $\underline{U}(x)$ can be treated directly as the "chromosomes" required by the genetic algorithm methodology. The information content of a permutation is bound up in its ordering of calls. By combining ordering information from two highly valued permutations (coding for some of the best solutions found so far), even better "offspring" permutations may be found. Ordering information in two permutations was combined by uniform order-based crossover, in which a random subset of elements in one permutation were reordered according to their ordering in the other permutation, as in the tutorial examples. Our second operator was scramble sublist mutation.

(b) Permutations are selected for combination *probabilistically*, according to their evaluations, using an exponential fitness technique. This technique has a single parameter p. It yields a sequence of fitnesses of the form $(1000, 1000 * p, 1000 * p * p, 1000 * p * p * p...)$. Given p equal to .9, for example, the fitnesses of population members would be (1000 900 810 729 ...). The value of p was interpolated over the course of our runs from .92 to .87, just as the value of the linear fitness technique is interpolated in the tutorial.

It is clear that hybrids of the preceding heuristics can also be formed. For example, the 2-opt improvement routine can be applied to the best permutation discovered by any other method in an effort to improve on it. However, this is a slow procedure, so it is desirable to use other methods first to get a good solution (close to the local optimum that 2-opt will end up climbing to). Otherwise, much time can be wasted by having the 2-opt procedure climb up (in a maximization problem) or down (in a minimization problem) from a starting point that is many steps away from the local optimum that is eventually reached. (A "step" in this setting is an improvement. Each improvement found marks the beginning of a new round of the time-consuming 2-opt improvement routine.) Comparison of the performance of the individual heuristics shows that the genetic algorithm performs better than the other algorithms over a wide range of computation time budgets, but that a hybrid strategy performs better still.

COMPARATIVE EVALUATION OF HEURISTICS FOR BANDWIDTH PACKING

Figure 11.1 shows how the different search heuristics performed on a 50-node test problem. To generate this figure, realistic link costs and capacities and a network topology for a 50-node network were assembled, based loosely on network data collected from the Denver and Colorado Springs geographic areas. The four heuristic approaches just described for solving the test problem were implemented and run to see how they compared in terms of quality of solution (the total profit of the best schedule discovered) as a function of CPU time invested. The results, averaged over many runs of each heuristic, are shown in the four curves in Figure 11.1. In this figure, the "2-opt" curve is seeded with the result of the greedy algorithm. Two alternative versions of the genetic algorithm are used—one that generates

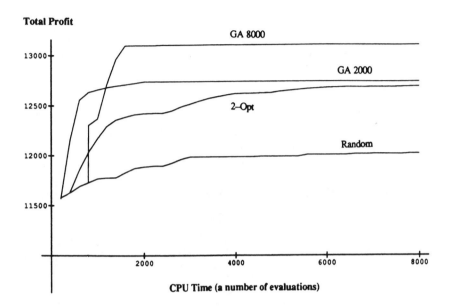

Figure 11.1: Comparison of algorithms for bandwidth-packing in a 50-node network.

a total of 2000 permutations and a slower one that generates a total of 8000 permutations before stopping. In Figure 11.1, these are called "GA 2000" and "GA 8000", respectively.

These results illustrate characteristics that we have confirmed in a variety of similar tests on this and other networks. The most important ones are as follows:

1. *All the heuristics, including uniform random search, do fairly well on these large problems.* (The uniform random search procedure achieves $12{,}019/13{,}097 = 92\%$ of the best solution found in this example.)

2. *All the heuristics can eventually get stuck at local optima:* none is guaranteed to find the global optimum. (For GA 8000, this was established by performing multiple runs and noting that the best value found varied slightly from run to run.)

3. *The curves are concave*, that is, the average additional improvement found per unit of CPU time diminishes with the total amount of CPU time invested, T.

4. *None of the four heuristics uniformly dominates the others*. Thus, for different time budgets T, different heuristics (or combinations of heuristics, e.g., random followed by 2-opt) are expected to make the greatest progress.

These conclusions were suggested by the 50-node test problem and by similar test problems with 10 nodes and 30 nodes. To verify them on a large set of randomly generated problems, we constructed a testing and evaluation platform for systematically evaluating the performance of bandwidth-packing heuristics under a range of task parameters. The test platform has facilities for (1) generating random "typical" communications network topologies; (2) generating increasingly heavy traffic loads on a given network topology; (3) simulating the arrival of call requests (traffic) for a given network topology under a given load pattern; (4) calling the bandwidth-packing heuristic being evaluated to have it route the offered traffic; and (5) collecting and displaying the performance statistics under different topology and traffic pattern (rate and mix) assumptions.

Table 11.1 and Table 11.2 present sample results obtained by different heuristics and hybrids of the pure heuristics on a 25-node test problem randomly generated by the platform. The first three columns describe the task parameters: the number of nodes and links in the network (25 and 36, respectively) and the number of call requests presented to be routed over it (varying from 25 to 45 in increments of 5). Other aspects of the network, such as link capacities and costs and average distances between nodes, as well as call request attribute values (bandwidth requirements and revenue if placed), were also randomly generated, for example, by uniform sampling from appropriate intervals, but are not shown in the tables. Column 4 of Table 11.1 shows the value of the path assignment determined by the greedy heuristic. It ranges from 92,528 when 25 calls are offered to 113,953 when 45 calls are offered. As more calls are offered, a lower percentage can be routed, but there is more opportunity to select profitable calls. The "Gr + 2" column shows the profit achieved by applying the opportunistic 2-opt improvement procedure to the permutation generated by the greedy heuristic, while the corresponding "steps" column records the number of improvements found by the 2-opt procedure in moving from its initial "seed" permutation (the one supplied by the greedy heuristic) to the best permutation that it produces. Since the 2-opt procedure is computationally very expensive compared to the other heuristics, the number of

Nodes	Links	Reqs	Greedy	Gr+2	Steps
25	36	25	92,528	101,271	5
25	36	30	98,028	106,130	7
25	36	35	101,763	115,898	16
25	36	40	110,585	123,365	23
25	36	45	113,953	127,644	22

Table 11.1: Results of Greedy and 2-Opt on Test Problems

Nodes	Links	Reqs	Initial GA	Final GA	Steps	GA+2	Steps
25	36	25	101,196	101,928	4	101,928	1
25	36	30	106,568	107,945	7	108,104	2
25	36	35	112,629	115,849	5	115,898	1
25	36	40	119,139	123,147	7	123,618	5
25	36	45	125,699	127,899	7	128,247	2

Table 11.2: Results of GA and 2-Opt on Test Problems

"steps" is a rough indicator of the relative amounts of CPU time required by different problems.

Table 11.2 presents analogous information for the GA 2000 heuristic. (Results for the random, random + 2-opt, and GA 8000 heuristics have also been collected for various problems but are in line with these results and are not reproduced here.) The "initial GA" and "final GA" columns show the profits discovered early in the genetic algorithm run (after the initial population of 200 random permutations was generated and evaluated but before controlled evolution began) and at the end of the genetic algorithm run (after a cumulative total of 2000 permutations had been generated), respectively. The "steps" column for the genetic algorithm shows the number of improvements, or "record values," in statistical terminology, discovered by the genetic algorithm in evolving from the initial to the final value. Finally, the last two columns present information on subsequent improvements found by applying the 2-opt improvement routine to the best permutation discovered by GA 2000.

The GA-followed-by-2-opt hybrid, denoted by "GA + 2" in Table 11.2, clearly achieves the best results. Not only are the values achieved higher than those for the other heuristics (including the uniform random search heuristic, which is not shown here), but also the run time for GA + 2 compares favorably to the run time for Gr + 2, the 2-opt procedure applied to a greedily generated seed. This may seem surprising, since the greedy heuristic is virtually instantaneous. The explanation is that, typically, the

GA 2000 component of the hybrid brings the search close enough to a local optimum so that the 2-opt part does not require great amounts of time to finish the climb. Metaphorically, 2-opt can be added to other heuristics to climb *further* than they do (going all the way to the local optimum, as judged by the 2-optimality criterion.) But GA 2000, which climbs toward multiple local optima simultaneously while it is operating, effectively climbs much *faster* than 2-opt can. Comparing the "steps" columns for Gr + 2 and GA + 2, we see that computational effort is economized by using a genetic algorithm to do most of the climbing, followed by 2-opt to finish the job.

For very short time budgets, random search may be the most effective strategy. However, the first part of the genetic algorithm consists precisely of random generation and evaluation of permutations; hence, the genetic algorithm can easily be reduced to random search for very short time budgets. This suggests the following three-way hybrid strategy. First, the greedy heuristic is run (at negligible time cost). Then, uniform random search is run to create and evaluate permutations until the desired initial population size for the genetic algorithm (e.g., 200 for GA 2000) has been reached. Next, the genetic algorithm is run until the desired cumulative number of individuals has been reached. Finally, 2-opt is applied until a local optimum is achieved or until the time budget T has been exhausted.

The empirical results presented in Table 11.1 and Table 11.2 are for a single, randomly generated, 25-node network. At this writing, the test platform is being used in a large-scale data collection effort. Over 100 randomly generated problems have already been run. The results in Figure 11.1 and these tables are not idiosyncratic: they are representative of results seen in other problems. Thus, the qualitative conclusions reported here appear to be sound.

The goal of on-going experimentation is to obtain quantitative information, for example, on the average improvement expected per additional unit of CPU time invested in a heuristic, which can be used to guide allocation of computational resources. Specifically, we are investigating the best initial population size and cumulative number of permutations to be used in the genetic algorithm for a given time budget. These two parameters determine when the proposed hybrid algorithm introduced in this chapter switches from random search to controlled evolution (the genetic algorithm) and when the genetic algorithm gives way to 2-opt.

REFERENCES

Kelly, F. P. (1986). Blocking probabilities in large circuit-switched networks. *Advances in Applied Probability* 18, pp. 473-505.

Kelly, F. P. (1988). Routing in circuit-switched networks: Optimization, shadow prices and decentralization. *Advances in Applied Probability* 20, pp. 112-144.

Meempat, G., and M. K. Sundareshan (1989). Adaptive bandwidth management for optimal access control of circuit switched traffic in integrated networks. Technical Paper, Department of Electrical and Computer Engineering, University of Arizona, Tucson, Arizona.

Roberts, J., and K. Liao (1985). Traffic models for telecommunication services with advance capacity reservation. *Computer Networks and ISDN Systems* 10, pp. 221-229.

Ross, K. W., and D. Tsang (1989). Optimal circuit access policies in an ISDN environment: A Markov decision approach. *IEEE Trans. on Comm.*, pp. 934-939.

Tsang, D., and K. W. Ross (1989). Algorithms to determine exact blocking probabilities for multirate tree networks. To appear in *IEEE Trans. Comm.*

12

A Genetic Algorithm Applied To Robot Trajectory Generation

Yuval Davidor

INTRODUCTION

This[1] chapter is concerned with two issues: the automation of robot trajectory generation, and the optimization of robot trajectories with a genetic algorithm. Robot trajectory generation belongs to the class of order-dependent processes, processes in which the order by which rules are executed has a fundamental effect on performance. Another distinctive feature of order-dependent processes is that they may be described by a variable number of rules. The generation of robot trajectories and the specification of robot trajectories form an archetype order-dependent process whereby valid robot trajectory programs may have any number (>2) of displacement commands. Robot trajectories can be generated automatically by a genetic algorithm, provided the algorithm will consider the order and

---

[1]The work presented here is based on an unpublished Ph.D. dissertation, *Genetic Algorithms for Order Dependent Processes Applied to Robot Path-Planning*, Imperial College, London (Davidor 1989a), and on a book, *Genetic Algorithms and Robotics: A Heuristic Strategy for Optimization* (Davidor 1990a).

varying lengths of trajectory specification programs.

This chapter describes a genetic algorithm that was designed to generate and optimize robot trajectories of any arbitrary system. The genetic algorithm for trajectory generation required extensive modifications to fit the inherent features of trajectories and the need to consider a variable number of rules and their order quality. The modifications are described in detail. The modifications made to accommodate variable-length and order-dependent strings are analyzed, and examples are given from the robot domain. The robot system used in this study is a redundant planner structure incorporating three links, but the model can easily be extended to an n-link structure. The performance of the genetic algorithm is compared with that of hill-climbing and random search algorithms, and shows characteristic improvements. The introduction of a sub-goal reward operator to the reproduction mechanism operator, based on Lamarckian probabilities for reproduction, improves the performance of the genetic algorithm.

BACKGROUND

Since the nineteenth century, as a result of the Industrial Revolution, machines have been employed more and more to gain economic affluence; the efforts to improve machine efficiency are continuing at an ever increasing rate. Almost every facet of our well-being in modern life involves machines that either produce or service our goods. The great efforts directed at improving the performance of production systems have engendered many optimizing techniques. However, the pace of front-edge technology becomes inadequate to meet the growing demands of our society, a society in which the complexity of manufacturing systems has reached such gigantic proportions as to become a bottleneck for manufacturing. What is needed is an efficient adaptive optimizing strategy that will be loosely dependent on prior knowledge of the solution.

The purpose of optimizers is to rectify performance when a deviation from a specified performance has been detected. Although the information on performance may be ample and easy to obtain, it becomes apparent that responding to this information is much more involved and difficult with redundant systems. Robot systems have frequently been equipped with redundant resources. The excess number of degrees of freedom creates a substantial problem for traditional control and optimizing strategies. The only way by which a classical optimizing technique can locate an optimum state is when either a sufficiently accurate model of the state-space

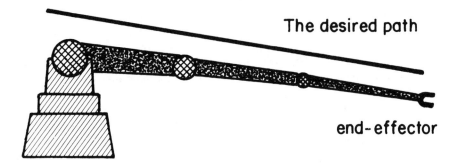

Figure 12.1: Three degrees of freedom planar robot arm at a fully stretched position, and a desired path for the end-effector to follow. Gravity and finite positioning accuracy result in the observed steady-state error from the desired horizontal position.

of the system exists or when the state-space is not too large, allowing a detailed investigation. Again, the interdependencies and the excess number of degrees of freedom preclude the effectiveness and reliability of traditional optimizing algorithms to locate the optimum state of operation. This is the environment into which an adaptive search strategy—a genetic algorithm— is introduced as a possible strategy to overcome the overwhelming complexity. The advantages of a genetic algorithm in such environments are the freedom from the need to possess an explicit model of behavior and the intrinsically parallel search dexterity which in redundant environments has the additional effect of distinguishing the important from the irrelevant, thus further simplifying the problem.

We should settle one important pragmatic issue at this early stage. Although, for reasons already discussed, the ideas developed herein use robot trajectory generation as an illustrative domain, it is not suggested that a several ton robot should train itself to follow a specified trajectory in arbitrary environments by waving its arm in a random fashion over some number of trials (no matter how few). Simulations and computer-aided design procedures, such as the one presented here, hedged with suitable safeguards, are what we need to help us realize the possibility of implementing these ideas.

ROBOT TRAJECTORY PLANNING

Most robot applications are based on a motion trajectory composed of a sequence of spatial displacements of a robot arm. Mechanically, a robot arm is an open kinematic chain composed of relatively stiff links with a joint between the adjacent links. Each link represents one degree of freedom and can be commanded to move independently of all other links. Standard systems have six degrees of freedom, to obtain full spatial flexibility. Since a robot arm performs a task through the motion of its *end-effector* attached to the last link, the last link is the primary component of the whole structure (Figure 12.1).

An *arm-configuration* is a unique arm structure defined by a set of positions of the links. Given these positions, the end-effector's position is computationally trivial and uniquely determined. Computing the end-effector's position from the links' positions is called direct kinematics. An end-effector trajectory is created by programming a sequence of spatial positions which the end-effector is required to visit between the starting and ending positions. Because only a finite number of positions can be specified and stored in any given program, the robot motion controller has to move the arm between the discrete intermediate positions following a dead reckoning procedure. Optimization of path-following programs means identifying the optimum combination and number of intermediate positions.

One can appreciate the complexity of programming a trajectory by examining the vertical plane in which the end-effector is required to follow the straight line connecting points A and B (Figure 12.2a). One robot program can list sites 1 and 2 for the end-effector to visit (Figure 12.2b) while another program considers the sites 3, 4, 5, and 6 to be more accurate (Figure 12.2c). Yet another possible trajectory, which also accounts for the effect of gravity, is composed of the sites 7, 8, 9, and 10 (Figure 12.2d).

Because of dynamic effects dependent on the hardware of the system, the quality of the resulting end-effector path of each of the alternative programs is quite different, although all trajectories aim to take the arm from A to B in a straight line. The theoretically equivocal question, how many intermediate positions are necessary or needed, is complicated by the fact that for redundant structures, end-effector positions have no unique arm structure and two trajectories have no clear homology in their defining arm-configurations. Furthermore, too often robot systems are assumed to have a predetermined performance. The practical ambiguity between specified and realized paths turns trajectory planning into a highly nonlinear problem. Although the above description of the system is minimal, it is possible to

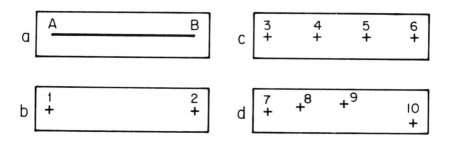

Figure 12.2: Illustrating the complexity in generating a trajectory specification. (a) The desired straight-line path between the starting end A and the terminating end B. (b) One possible trajectory specified by two arm-configurations that correspond to the end-effector being at the two ends of the desired path. (c) Another alternative trajectory with two intermediate end-effector positions along the desired path. (d) Another possible trajectory specification for the straight-line path.

see that the mechanical inaccuracies, mechanical coupling, and dynamic effects turn trajectory generation into a complex problem to optimize.

APPLYING A GENETIC ALGORITHM TO ROBOT TRAJECTORY GENERATION

The problem domain of robot trajectory generation exhibits features indicating that a genetic algorithm might be a rewarding optimizing tool to use in this domain. However, it is impossible to apply any of the existing genetic algorithm models because they lack the ability to manipulate variable-length strings required by the natural representation of robot trajectories. A new genetic algorithm model has to be developed to optimize robot trajectories.

The Representation of a Trajectory

The set of link positions that define an arm-configuration is the basic string building block in this discussion. The decision regarding this representation structure was made primarily because it coincides with the natural

n-tuple

$$\alpha_{1,1};\ \alpha_{1,2};\ ...;\ \alpha_{1,n};\ \alpha_{2,1};\ \alpha_{2,2};\ \alpha_{2,3}\ ...\ ;\ \alpha_{l,n}$$

l n-tuples

Figure 12.3: The string structure of a trajectory. The string is composed of angles, $\alpha_{i,j}$, which are divided into n-tuples (the index i denotes the position of the ith link in the jth arm-configuration). Each n-tuple represents one arm-configuration, and a total of l n-tuples is used to specify a given trajectory of l arm-configurations.

structure of trajectories, a sequence of end-effector positions defined by the corresponding arm-configurations. It is now possible to construct the full string (see Figure 12.3). A string is composed of link positions (angles) grouped in n-tuples. Each n-tuple represents a single arm-configuration, where the first angle in the tuple corresponds to the position of the first link and so forth for all n links in the arm. In a way, the basic unit of the string is the n-tuple which repeats l times to define the l arm-configurations used to define a given trajectory. Hence, an l-trajectory is a trajectory defined by l arm-configurations. In order to make the robot model interesting, a redundant arm structure was used (a planar 3-link robot arm as illustrated in Figure 12.4). This means that for a particular end-effector position, a multitude of valid arm-configurations produce the given end-effector position. (Details on the size of the trajectory space are given in section "Some Trajectory Generation Trails".) Without going into details about the dynamic equations of the robot simulation, it will be noted that all dynamic effects, besides link elasticity, are considered: gravity, centrifugal and coriolis forces, backlash in the gears, dynamic and static friction of the actuators, and delays in the control system.

The trajectory space is clearly a multidimensional, multimodal space. The high degree of multimodality can be appreciated when the redundant link is remembered. If the robot is entrusted with a specific path to follow, then for a fixed position of one of the links (apart from a few singular positions), there is a family of valid trajectories called the root family of the given link position. Each root family can be further divided into subfamilies depending on the position of the second link.

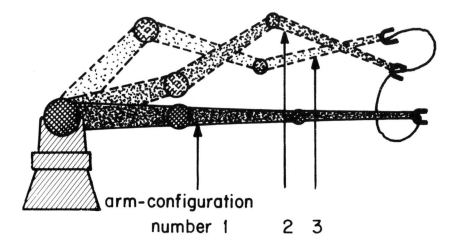

Figure 12.4: *3*-trajectory with the three arm-configurations, and the produced end-effector's path.

Why Varying-Length Representation?

As demonstrated with the simple straight-line path, robot trajectories must consider trajectories of a varying number of arm-configurations. A variable number of arm-configurations are mapped in the representation scheme discussed above into a variable-length string of link angles. Most genetic algorithm models are not equipped to handle variable-length strings. For example, any simple recombination scheme, attempting to recombine trajectory (b) with trajectory (c), is faced with a serious compatibility dilemma: how to recombine the two and yet produce a sensible trajectory.

The recombination dilemma is further complicated by adding the order quality of the arm-configurations. It makes little sense to construct a trajectory from sites 9, 8, 7, and 10 (Figure 12.2). It is clear that recombining trajectories *b*, *c*, or *d* through the conventional crossover operators will produce too many meaningless offspring trajectories. Specialized recombination schemes have been suggested before (see Goldberg 1985, Grefenstette 1985, Schaffer 1987, and Smith 1980), and at least 56 different crossover operators are documented. However, both for applications requiring a varying number of string elements which do not conform to a straightforward homologous structure, and for a more application-free genetic algorithm, the issue of varying-length strings has to be approached with a more general representation model and algorithm (Davidor 1989b).

Goldberg recently introduced a notable type of a genetic algorithm—a messy GA (Goldberg 1989)—which processes variable-length strings that may be either under- or over-specified with respect to the problem being solved. Goldberg realized that the robust and adaptive quality which natural selection has in respect to information processing is the ability to process overspecified chromosomes. He describes this feature of natural selection as "nature's climb out of the primordium", the adaptation of simple organisms that paves the way for more sophisticated organisms to appear. The messy genetic algorithm represents a fundamental change in the approach to adaptation and is in its very first stages of development.

The most central issue in this chapter is the use of variable-string lengths and the order-dependent representation. Smith's Learning System LS-1 (Smith 1980) proposed to overcome the variability in length by incorporating multilevel reproduction operators. Reproduction in the presence of variable-length strings necessitates modifications of all reproduction operators, especially for crossover. Other variable-length representations were presented by Schaffer (1984) in his LS-2 learning system, by Cramer (1985) to manipulate sequential tree-structured programs, by Bickel (1987) to create lists of tree-structured production rules, by Greene and Smith (1987) to learn models of consumer choice, and by Fujiki and Dickinson (1987) to solve the prisoner's dilemma. Not all of these models addressed the order-dependency issue and so not all apply to the robot paradigm.

The reproduction operator

The main modifications which conventional genetic algorithms require in order to accommodate variable-length, order-dependent strings occur in the recombination operator. In traditional recombination operators, after choosing a cross site in one parent string, the corresponding cross site in the second parent is readily determined according to the locus of the cross site in the first parent. Hence, corresponding cross sites are situated in the same positions in their respective strings, and string position is the criterion for the match. This type of cross site matching criterion can be associated with the biological term *homology*[2]. However, homologous cross sites are

---

[2]The use of the word *homology* here refers to the fact that conventional string representations have corresponding structures when crossed. This contrasts with the biological use, in which homologous organs are body organs of different species sharing the same evolutionary origin that have come to serve different purposes. The members of each pair of chromosomes are called homologous chromosomes and exhibit a specific

not suitable for variable-length, order-dependent strings owing to the weak correlation between the parameter's relative position in a string and their resulting effect on the phenotype. The variable-length and order character of trajectories nullifies the role position has in fixed-length representations. The order dependency prevents the use of scaling operators to normalize the string length (such as the scaling effect of the dynamic time warping used in speech recognition).

The variable-length, order-dependent strings use an *analogous* criterion for matching cross sites—a cross point correspondence criterion based on similarity of the phenotypic function of the parameter at the cross site locus. The crossover operator is therefore called *analogous crossover* (1989b). Matching parameters for crossover according to their phenotypic function, rather than according to their genotypic position, is a fundamental modification of the recombination operator. After choosing a cross site in one parent string, the corresponding cross site in the second parent is determined according to the proximity the parameters have in the phenotype space. The locus of the cross site of the first parent has neither significance nor a role in determining the locus of the corresponding cross site in the second parent.

Matching parameters according to their phenotypic function can be illustrated with the road system. Driving between two arbitrary points A and B, we can devise several alternative routes, while some routes may even have shared stretches. The logical and economical way of combining different routes is to switch over at either shared stretches of road or at intersections of routes. Nevertheless, the cavalier driver might venture switching to routes when he is not close to a shared stretch (or to a route that has no shared stretches with his current route) and, thus, suffers the additional travel distance due to the connection. For that reason, switching routes is more likely to occur at route positions where the two routes are close to each other (Figure 12.5).

The analogous cross site criterion not only preserves the "orderliness" characteristic of strings, but also has a biological motivation and intuitive justification (as demonstrated with the road travel example). In a complex string structure where the number, size, and position of parameters have no rigid structure, it is important that crossover occurs between sites that control the same, or at least similar, function (in the phenotype space). The extent of variability in length of strings owing to the modified representation, which caused the precise locus to have a diminished significance,

---

length, centromere placement, and identical gene sites (or loci) along their axes. Similarly, *analogous* organs are organs or parts of the body of different species that have different evolutionary origins but have become adapted to serve the same purpose.

path 1

path 2

Figure 12.5: Two paths resulting from different trajectories. The circled sites are preferable for cross point sites.

makes a mechanism like the analogous crossover a necessity.

One should be careful to interpret the analogous recombination as a monotonically successful crossover. First, the decision on the two cross sites in the first parent is random (that will change once Lamarckian heredity is introduced in the next section), paying no attention to what are the resulting cross sites, or with what second parent trajectory the recombination is going to happen. Second, the matching criterion used here is the proximity of the discrete end-effector positions. This criterion may be misleading because of the redundant link (as in the case of assuming a match between the two end-effector locations in Figure 12.6). Matching end-effector proximity is not the only plausible matching criterion. For example, proximity in the net links position may serve as an alternative criterion. Proximity in the end-effector space promoted a more harmonious crossover and therefore served as the matching criterion in this application.

A subtle feature of the crossover operation—a feature that was not visible when only single-bit parameters were used—becomes relevant in the presence of multi-bit parameters (the n-tuples). In order to differentiate between the two alternative cross sites (inside a n-tuple and outside it), the crossover operation that crosses outside n-tuples is called a *segregation crossover*. When multi-bit parameters have a natural meaning, crossing over within those groups may have a disruptive effect on the string structure and consequently on the phenotype function (Green and Smith 1981; Smith 1980).

Segregation recombination proceeds according to the following sequence:

1. Choose two strings.

2. Randomly choose an n-tuple (arm-configuration) in one string.

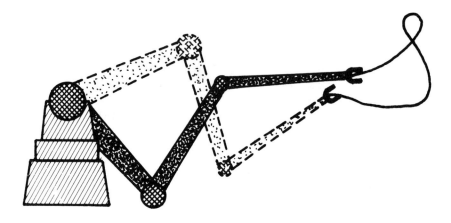

Figure 12.6: A vigorous arm movement is encountered, although the two end-effectors are positioned in close proximity. This vigorous movement is the result of a motion between two very different arm-configurations.

3. Slide the second string from head to tail along the first in order to find its most analogous arm-configuration to the one chosen in the first string. The cross site will be after the n-tuple representing the respective arm-configurations.

4. Repeat steps 1 and 2 for the second cross site.

5. Copy the counterpart sub-strings from the two parents to obtain a new offspring string.

From the above two paths it is obvious that it is attractive to cross over at the regions where the two paths meet (Figure 12.5). However, in order to cross over at phenotypically similar sites, the genotype must contain the phenotype description, or a transmission function from genotype to phenotype, two conditions that are alien to most biological systems and conventional genetic algorithms. Alternatively, analogous matching can be processed on the basis of genotype information, such as in assembly applications where each step is distinctively identified. In the robot paths application, the analogous cross sites are matched in the genotype space on the basis of the proximity of the end-effector position.

Figure 12.7 shows a compressed version of two trajectory genotypes (compressed in the sense that only the discrete end-effector positions are

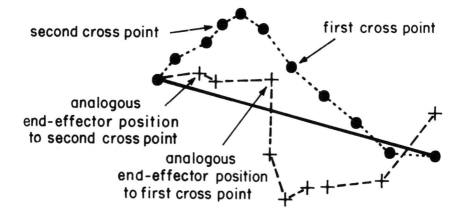

Figure 12.7: Two end-effector trajectories, the two cross sites of one string, and the two analogous matched cross sites in the second string.

displayed for each of the arm-configurations composing the two trajectories). The two randomly chosen cross sites are marked in one of the parents (solid circles), whereas the analogous matching points are shown in the second parent (crosses). The crossover is a "crossing after" operator. Note that the resulting offspring trajectory is different from its parent trajectories in the number of arm-configurations that comprise it (Figure 12.8).

The recombination operator intrinsically involves the risk of losing important schemata through the premature loss of string diversity. Although string diversity in fixed representations means the loss of diversity at certain string positions, in the variable-length representations it also means the possible loss of diversity in the number of arm-configurations defining the trajectories. In the varying-string-length environment, the mutation operator is expanded to introduce random changes to the length of strings. The two additional mutation operations, shortening and lengthening strings, are called *deletion* and *addition*, respectively. These two operators operate as a background noise over the strings' length. The deletion operator simply deletes an n-tuple (one complete arm-configuration), whereas the addition operator duplicates an arm-configuration and keeps the two adjacent to each other. The effect of the addition operation becomes clearer when we consider the combined effect it has with crossover, especially in the case of analogous crossover. The purpose of the addition operator is

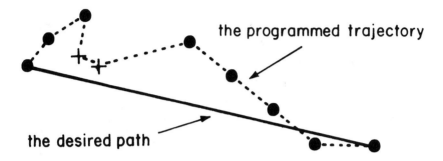

Figure 12.8: Offspring trajectory resulting from crossover of the two trajectories shown in Figure 12.7.

twofold: to enrich the pool of arm-configurations so that crossover can occur at more sites; and to promote the proliferation and propagation of arm-configurations throughout the arm-configuration pool in the population. The effect of the addition and deletion operators is subtle and can be better appreciated when Lamarckian probabilities for reproduction are introduced, or in a more detailed discussion of this model (1990a).

LAMARCKIAN PROBABILITIES FOR REPRODUCTION

The Lamarckian theory of the inheritance of acquired characteristics lost its followers when modern research did not support the principal arguments of the theory. Though infrequent discoveries bring an occasional renaissance to the Lamarckian theory (Berek 1985; Gorezynski and Steele 1981), it is generally accepted that it is not feasible to inherit acquired qualities. This is the major reason why many algorithms, simulating or mimicking genetic processes, have ignored the Lamarckian theory and avoided implementing its ideas. From the machine learning viewpoint, Lamarckism is an important model that can complement many learning algorithms.

Several applications have addressed the possibility of improving learning by incorporating operators of a Lamarckian nature. The validity of

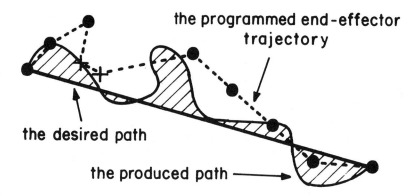

Figure 12.9: The produced path and the accumulated error from the desired path.

the Lamarckian inheritance of acquired characteristics in natural systems is debatable, but various mechanisms have been suggested in which such directed learning can enhance machine learning. The most relevant models are ARGOT (Shaefer 1987), VEGA (Schaffer 1984), and the adaptive crossover distribution genetic algorithm (Schaffer 1987).

The nature of sequential processes makes the application of sub-goal reward schemes plausible in this environment (Davidor 1990c). An example is the assignment of the proportional error contribution of each of the arm-configurations used in a particular trajectory. If the goal function aims to minimize the accumulated error from the desired path (integration of the error in Figure 12.9), then it is possible to approximate the relative error contribution that any displacement between two arm-configurations had (Figure 12.10). However, sub-goal reward, which is greatly dependent on the amount of epistasis in the representation structure, cannot always be assigned in a straightforward way (Wilson 1987).

Obviously, the quality of a sub-goal fitness prediction is dependent on the degree of epistasis present in the representation; the less epistasis among the string parameters, the more accurate the prediction can be (Davidor, 1989c). In the robot trajectories application, performance is greatly dependent on the relations between adjacent arm-configurations. An unharmonious displacement between two arm-configurations which results in a relatively large error (like the displacement in Figure 12.6), indicates that

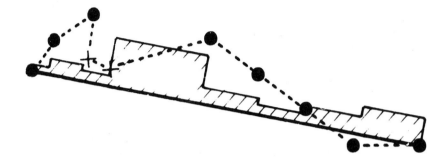

Figure 12.10: Normalized and digitized accumulated error distribution as a function of the arm-configuration sequence.

the two are not relatively co-adapted. If a sequence of arm-configurations has a relatively low error contribution, then this sequence is relatively more co-adapted than other sequences in the trajectory.

This kind of sub-goal information can assist the reproduction operators to predict where it is more promising to operate and where to concentrate search efforts. It is true that the sub-goal fitness estimate is just an estimate and may be misleading. The schema theorem indicates that it is advantageous to cross over on the borderline between coadaptation of a schema. The effect of the inversion operator is to encourage co-adapted parameters to be closely positioned on the string to minimize the probability of separating them through crossover. Although it is impossible to use the inversion operator here in a straightforward fashion (owing to the order dependency), we can regain its effect by altering the probability of choosing cross sites according to the error distribution along the string (Schaffer 1987). This kind of Lamarckian inheritance has a positive effect on the efficiency and stability of the algorithm. Replacing the uniformly distributed mutations with a probability distribution based on the sub-goal fitness has a significant effect on the efficiency of the mutation operator. Using probability-distributed mutations preserves the benefits of the mutation operator and diminishes the counterproductive effects it has in destroying good schema. By concentrating the mutation efforts at loci that are not co-adapted, or that have a relatively low fitness contribution to the global fitness of the organism (Figure 12.11), the mutation operator

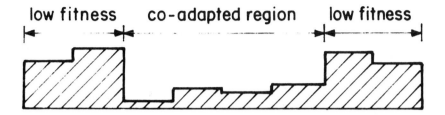

Figure 12.11: A Lamarckian probabilities distribution that is used for all the recombination operators.

introduces new information to the population but does not tend to destroy good information. The Lamarckian probability for mutations enables the mutation operator to be more controlled and to introduce some qualities of a local hill-climbing operator.

SOME TRAJECTORY GENERATION TRAILS

At this stage, after discussing the special features of the order-dependent processes and the robot trajectory generation application, it is appropriate to sketch the behavior of the modified algorithm. The task is to optimize a path-following program of a 4.2 m long straight-line path (the desired path in Figure 12.1). The size of the trajectory space that is considered for this application is in the region of $\approx 10^{40}$. The population size is set to be 100. The results presented are typical convergence results. (The detailed experimental data are considered to be beyond the scope of the present discussion but can be found elsewhere—see Davidor 1990a.) Figures 12.12 and 12.13 plot the convergence rates for the three search algorithms: random search, hill climbing, and the genetic algorithm. These results clearly indicate that our trajectory generation domain is complex and multimodal, as was suggested earlier.

The genetic algorithm for robot trajectory generation was applied to a set of trials twice: once with reproduction based on uniform probability, and once on a Lamarckian probability. The comparative study between the two experiment conditions is summarized in Figure 12.14 and covers a total of 20,000 trials. It should be remembered that the Lamarckian proba-

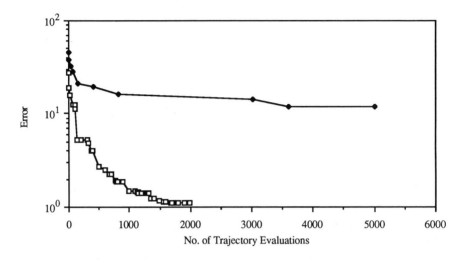

Figure 12.12: Simulation results of a random search (solid diamonds) and a genetic algorithm (hollow squares).

bility represents a subtle change in the way the genetic algorithm operates. Instead of choosing an arm-configuration as a cross site with equal probability, the arm-configurations are chosen probabilistically according to their respective estimated local error. What gains should we expect from thus changing the probability on which the reproduction operators base their operative decisions? The effect of the Lamarckian probability on recombination was discussed in general terms in reference to sub-goal reward. For genetic algorithms in general, Lamarckian probability (as defined here) has a beneficial and noticeable effect on performance when the population exhibits two features:

1. If the correlation between the local error distribution and true performance is fairly good.

2. If the error distribution along a string is significantly nonuniform.

The meaning and applicability of the first condition were discussed earlier. The second condition requires a large diversity of error distribution along the trajectories because, otherwise, the Lamarckian probability degenerates. It has a similar effect to that of a uniform probability

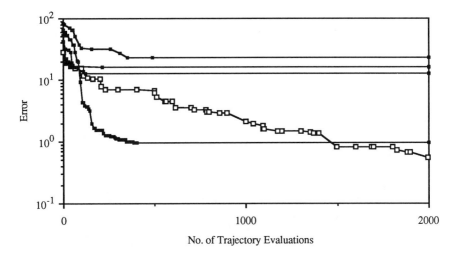

Figure 12.13: Simulation results of four runs of a hill-climbing algorithm (solid squares), and a genetic algorithm (hollow squares).

crossover. As a result of the above two conditions, the Lamarckian probability should exhibit significantly different performance only at search stages where trajectories are unharmonious and the diversity of co-adapted arm-configuration sequences is sufficiently large.

A large number of generations is required before the effect of the Lamarckian probability distribution can be clearly observed. The limited number of overall trials generated at each experiment reduces the relative efficiency of the Lamarckian probability over uniform probability. The last factor that lessens the efficiency of the Lamarckian probability over the uniform one in this model is the relatively small number of arm-configurations that compose trajectories. The smaller the number of arm-configurations, the fewer the possible number of co-adapted sequences in a string. The fewer co-adapted sequences a trajectory can include, the closer the error distribution is to a uniform one—an adverse situation when applying Lamarckian probability.

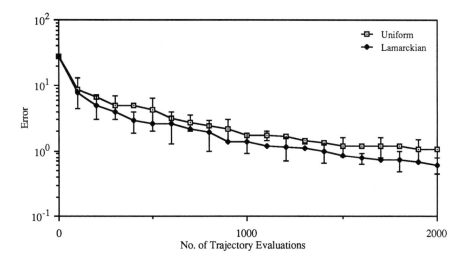

Figure 12.14: Statistical summary of five pair experiments comparing the performance of a genetic algorithm with uniform probabilities for reproduction operators (solid squares) with that of a genetic algorithm with Lamarckian probabilities (hollow diamonds).

CONCLUSIONS

Order-dependent processes emerged from the broad band of manufacturing as a conceptual trend in the analysis of industrial processes. Although they raise substantial obstacles for implementing traditional optimizing methodologies, appropriately tailored genetic algorithms can be applied successfully. The main purpose of this chapter was to present a genetic algorithm model that is capable of processing order-dependent, varying-in-length representations suitable for describing and optimizing a typical order-dependent process—the generation of robot trajectories. Such an algorithm was developed and proved to be successful.

Robot trajectories can be represented in their natural format. The analogous crossover can direct crossover operations in variable-length string structures while maintaining order in the order-dependent structures. The effect of Lamarckian probability could not be fully analyzed because the conditions required to examine these effects were in conflict with the other requirements of the particular application. However, even under these unfavorable conditions, the Lamarckian probability exhibits significantly improved performance over the uniform probability scheme.

That this model was neither developed for a particular robot system nor uses any of the sophisticated procedures which contemporary industrial robot systems offer suggests that the performance and utility of this application may be substantially increased with further development. This possibility is presently being investigated by the author at the Weizmann Institute of Science.

ACKNOWLEDGMENTS

The author wishes to thank the Centre for Robotics and the Department of Computing at Imperial College where most of this work was carried out; Dr. Antonia J. Jones and Dr. Tom H. Westerdale with whom the author enjoyed many fruitful discussions and received support; and Yehuda Barbut for producing the art work for the drawings.

A special thanks to Dr. Leon Zlajpah of Jozef Stefan Institute for allowing the free use of the robot dynamic simulation and for active support in tailoring the simulation to suit the special requirements set by the thesis.

REFERENCES

Berek, C., et al. (1985). Molecular events during maturation of the immune response to oxazolone. *Nature*, 316, pp. 412–418.

Bickel, A. S., and Bickel, R. W. (1987). Tree structured rules in genetic algorithms. *Second International Conference on Genetic Algorithms and their Applications.*

Cramer, N. L. (1985). A representation for the adaptive generation of simple sequential programs. *First International Conference on Genetic Algorithms.*

Davidor, Y. (1989a). *Genetic algorithms for order dependent processes applied to robot path-planning*, Unpublished Ph.D. dissertation, Imperial College, University of London.

Davidor, Y. (1989b). Analogous crossover. *Third International Conference on Genetic Algorithms.*

Davidor, Y. (1989c). Epistasis variance—suitability of a representation to genetic algorithms. *TR–CS89–25*, Weizmann Institute of Science.

Davidor, Y. (1990a) *Genetic Algorithms And Robotics: A Heuristic Strategy for Optimization*. Singapore: World Scientific Publishing Co.

Davidor, Y. (1990b). Robot programming with a genetic algorithm. *IEEE International Conference Proceedings on Computer Systems and Software Engineering.* Tel-Aviv: Israel.

Davidor, Y. (1990c). Sub-goal reward and Lamarckism in a genetic algorithm. *Proceedings of the European Conference on Artificial Intelligence.* Stockholm: Sweden.

Fujiki, C. and Dickinson, J. (1987). Using the genetic algorithm to generate LISP source code to solve the prisoner's dilemma. *Proceedings of the First International Conference on Genetic Algorithms and their Applications.*

Goldberg, D. E. (1987). Alleles, loci, and the traveling salesman problem. *Proceedings of the First International Conference on Genetic Algorithms and Their Applications.*

Goldberg, D. E. (1989). Messy genetic algorithms: Motivation, analysis, and first results. *TCGA Report No. 89003*, Department of Engineering Mechanics, University of Alabama, Tuscaloosa, Ala 35487–2908.

Gorczynski, R. M., and Steele, E. J. (1981). Simultaneous yet independent inheritance of somatically acquired tolerance to two distinct H-2 antigenic haplotype determinants in mice. *Nature* 289, pp. 678–681.

Greene, D.P., and Smith, S. F. (1987). A genetic system for learning models of consumer choice. *Second International Conference on Genetic Algorithms.*

Grefenstette, J. J., Gopal, R., Romaita, B. J., and Van Gucht, D. (1985). Genetic algorithms for the traveling salesman problem. *First International Conference on Genetic Algorithms and their Applications.*

Holland, J. H. (1975). *Adaptation in Natural and Artificial Systems*. The University of Michigan Press.

Shaffer, J. D. (1984). *Some Experiments in Machine Learning Using Vector Evaluated Genetic Algorithms*. Unpublished Ph.D. dissertation, Vanderbilt University, Nashville, Tennessee.

Schaffer, J. D. and Morishima, A. (1987). An adaptive crossover distribution mechanism for genetic algorithms. *Second International Conference on Genetic Algorithms*.

Shaefer, C. G. (1987). The ARGOT strategy: Adaptive representation genetic optimizer technique. *Second International Conference on Genetic Algorithms*.

Smith, S. F. (1980). *A Learning System Based on Genetic Adaptive Algorithms*. Unpublished Ph.D. dissertation, University of Pittsburgh.

Wilson, S. W. (1987). Hierarchical credit allocation in a classifier system. In L. Davis (ed.), *Genetic Algorithms and Simulated Annealing*. London: Pitman, pp. 104–115.

13

Genetic Algorithms, Nonlinear Dynamical Systems, and Models of International Security

Stephanie Forrest and
Gottfried Mayer-Kress

INTRODUCTION

Nonlinear dynamical systems is an area of active research (see, for example, Eckmann and Ruelle 1983). A nonlinear dynamical system typically consists of a set of differential equations (or difference equations if time is discrete) that model how some system changes over time. Equation 13.1 in the section on Richardson models defines one such system. The long-term behavior of dynamical systems can be difficult to predict analytically because of nonlinear couplings between dependent variables. Consequently, an empirical approach is commonly used to understand their behavior. Under this approach, the equations are iterated for long periods of time with different initial conditions to determine asymptotic behavior (such as fixed points, periodic or chaotic oscillations, etc.) and stability to noise. Even for simple models, the constant parameters of the equations can have a tremendous effect on their long-term behavior.

One limitation of the computational approach to studying nonlinear dynamical systems is that for realistic models the number and range of possible parameter settings can be enormous. For example, in Equation 13.1 there are twelve real-valued parameters: x_s, y_s, z_s, x_m, y_m, z_m, k_{11}, k_{22}, k_{33}, k_{23}, k_{13}, and k_{12}. In such cases, the parameter space is so large that it is infeasible to search it exhaustively. A single point in the parameter space is expensive to evaluate. Each equation of the system is iterated until it reaches asymptotic behavior under many different initial conditions, and the system is tested for sensitivity by seeing whether small changes of initial conditions lead to qualitatively different asymptotic behavior. Genetic algorithms offer the possibility of searching the parameter space intelligently to find regions of interest. This chapter describes an application of genetic algorithms in which the problem of finding interesting regions of parameter space is reduced to a nonlinear optimization problem.

Historically, the computational approach to dynamical systems has been used to study the behavior of physical systems (fluid flows, convection, weather patterns, etc.), but recently there has been increasing interest in modeling social phenomena, such as stock markets and arms races (Grossman and Mayer-Kress 1989, Mayer-Kress 1989, Saperstein and Mayer-Kress 1988, Farmer 1989). The dynamical system presented in this chapter is an arms-race model. It describes how expenditures for arms change over time in response to previous expenditures and perceived external threats. In this context, the genetic algorithm also provides a tool for studying how policies evolve over time. In models of social phenomena such as the one described here, the dynamical systems approach makes it possible to study the long-term implications of various policy decisions. In the real world, however, policies are rarely fixed for long periods of time. Rather, they are changing on the same time scale as expenditure levels, that is, potentially at every time step of the simulation. Thus, the genetic algorithm plays two roles in the following: (1) a nonlinear optimization procedure, and (2) a model of evolution and negotiation.

RELATED WORK

The work reported here is based on two previous research efforts. Nonlinear dynamical systems theory has been used to study various models of strategic interactions, particularly arms-race models (Saperstein and Mayer-Kress 1988, Grossman and Mayer-Kress 1989, Mayer-Kress 1989). The goal of this work is to understand the long-term implications of fixed policies on global stability. In particular, we use an extension of the two-party Richardson model (Richardson 1960a, Richardson 1960b) to three agents. Axelrod and Forrest used genetic algorithms to study the evolution of cooperative strategies in the Prisoner's Dilemma and Norms Game (Axelrod 1987, Axelrod 1986). In that work, each individual had its own policy (strategy), and the genetic algorithm was used to evolve policies over time. These projects provide important background for the current work and are reviewed briefly in the following two subsections.

The Richardson Model

The Richardson model is a classic model of two-party arms races (Richardson 1960a, Richardson 1960b). In this model, individual agents represent countries, and the parameter settings represent specific armament policies. The model describes how each country's spending on arms changes over time, as a consequence of its own and the other countries' previous spending. In studying this class of models, the primary objective is to understand how policies could have predictable effects over long time-scales.

Natural extensions of the Richardson model can be used for multiple-agent interactions, as described in Mayer-Kress (1989). In the three-agent Richardson's model the agents are the countries X, Y, and Z. The two weakest countries are always allied against the strongest, where strength is determined by current expenditures on arms. An important form of instability in this type of model is a shift in alliance (e.g., if Y and Z are allied against X, and Y becomes the strongest nation). The model consists of the following three equations:

$$
\begin{aligned}
x_{t+1} &= x_t + (k_{11}(x_s - x_t) + k_{23}(y_t + z_t))(x_m - x_t) \\
y_{t+1} &= y_t + (k_{22}(y_s - y_t) + k_{13}(x_t - z_t))(y_m - y_t) \qquad (13.1) \\
z_{t+1} &= z_t + (k_{33}(z_s - z_t) + k_{12}(x_t - y_t))(z_m - z_t)
\end{aligned}
$$

x_t, y_t, and z_t represent the arms expenditures of the three countries for the previous year. x_s, y_s, and z_s specify the intrinsic arms expenditures of each country—that is, how much each country spends on arms independently of the spending by other countries. The terms $(y_t + z_t)$, $(x_t - z_t)$, and $(x_t - y_t)$ denote the external threat from adversaries for the countries

X, Y, and Z, respectively. Equation 13.1 is written with the assumption that X is the dominant country (most money spent on arms), so that Y and Z are allied against X. Thus, the threat to X is the sum of Y's and Z's spending on arms $(y_t + z_t)$. If a country is in an alliance, then its threat is reduced by the amount that its ally spends (e.g., $(x_t - y_t)$ represents Z's external threat). In the case where Y or Z becomes dominant, these terms are modified to reflect the correct external threat. x_m, y_m, and z_m represent the economic constraints on X, Y, and Z, that is, what fraction of the country's resources are available for spending on arms. For convenience, we assume that each country has the same total wealth, so that x_m, y_m, and z_m denote the relative resources that each country has available for arms at each time step. Finally, the terms k_{11}, k_{22}, k_{33}, k_{12}, k_{23}, and k_{13} are rate constants. Roughly, they control the "defense intensity," or how quickly a country responds to an increase or decrease in external threats.

Each country's policy is determined from a set of parameters. By iterating the equations for many time steps, it is possible to observe what the steady-state behavior of the model will be under different policies. Varying the initial conditions, parameter settings, and noise allows us to study the long-term effects of various policies. Instabilities can arise from any of these factors. Of particular interest are shifts in alliances and unlimited arms races. For example, some scenarios will lead to unstable behavior in which the variables grow without bound (an arms race), while others will lead to stable fixed points (arms control).

Figure 13.1 illustrates instabilities that arise as a result of noise in this model. The figure shows a two-dimensional slice through the 12-dimensional parameter space of the model: k_{11} is varied along the X-axis, z_s is varied along the Y-axis, and the remaining parameters are set to standard values. Each point in a subfigure shows the relative strength of the different countries after $T = 33$ time steps (years): bright = X is strongest (i.e., $x_t > y_t$ and $x_t > z_t$), medium = Y is strongest, dark = Z is strongest. Each axis is divided into 64 different values, so each subfigure shows the outcome of $64 \times 64 = 4096$ time histories. The boundary between regions of different shades of grey corresponds to a change in the alliance configuration, a case that we associate with a major crisis. The subfigures show the effect of different noise levels (σ = perturbation of the system parameters averaged over $N = 1000$ time histories): (a) no noise $\sigma = 0$, (b) $\sigma = 0.001$, (c) $\sigma = 0.003$, (d) $\sigma = 0.005$. Noise is added to the system at each time step n by replacing the original system parameters k_i by $k_i(n)$ = $k_i + \xi_n$ where ξ_n is a random number uniformly distributed in $[-\sigma, \sigma]$. The figure shows smooth parameter boundaries when there is no external noise and increasingly irregular boundaries (regions of alliance shifts) as the noise level increases. This implies that the model's behavior is much more difficult to predict in (d) than in (a).

Experimentation with the three-agent Richardson model and its relatives has demonstrated conditions under which the model leads to instability and stability. For example, the "defense intensity" parameters have a threshold above which the system tends to unlimited arms races and below which a stable balance of power can be established. Typical parameters and initial conditions lead to multiple attractors and high sensitivity to noise. This implies that the model is effectively unpredictable in some regions (Mayer-Kress 1989).

Genetic Algorithms and Cooperative Behavior

Axelrod and Forrest used genetic algorithms to study the evolution of cooperative strategies in game-theoretic environments, including the Prisoner's Dilemma and an n-agent game called Norms (Axelrod 1987, Axelrod 1986). The Prisoner's Dilemma is a well-known two-agent game, defined by a payoff matrix, an example of which is shown in Figure 13.2 (Tucker 1950). In the Prisoner's Dilemma each player has two options—cooperation or defection. The objective of the game is for each agent to maximize its score. The Prisoner's Dilemma is a *non-zero sum* game because the total number of points is not identical in all four boxes. Collectively, the players do best if both cooperate, with a total of six points, even though each player can potentially do better individually by defecting. For one round of the Prisoner's Dilemma, it is straightforward to show that each player is best off defecting. However, for iterated Prisoner's Dilemmas, in which the same set of opponents plays the Prisoner's Dilemma repeatedly, the best strategy is not so obvious. A *cooperative strategy* tries to establish a pattern of mutual cooperation with its opponent, yielding the greatest total of points for the group. An *exploitative strategy* attempts to defect while its opponent cooperates, potentially resulting in the greatest number of individual points. Since the strategies are designed to be completely self-interested, that is, there is no altruism, an important question is if and how cooperative behavior can evolve from groups of self-interested agents (Axelrod 1984).

The genetic algorithm was used to study how individual self-interested strategies evolve over time. Each individual in the population defined a strategy for playing the Prisoner's Dilemma, and the fitness function measured the number of points scored in an iterated Prisoner's Dilemma of 151 moves. In some cases, populations were evolved against a fixed set of strategies (each individual's fitness was determined by playing a fixed set

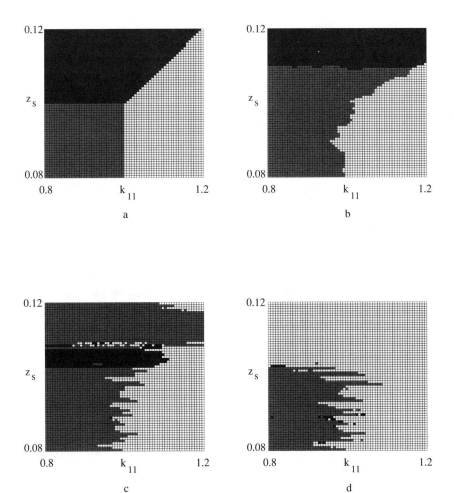

Figure 13.1: Effect of noise perturbation on typical parameter settings.

	B Coops	B Defects
A Coops	(3,3)	(0,5)
A Defects	(5,0)	(1,1)

Figure 13.2: Prisoner's Dilemma payoff matrix. The payoffs for Players A and B are specified for each of the four possible outcomes of a one-shot Prisoner's Dilemma (A defects and B cooperates, A cooperates and B cooperates, etc.). The first score in each box is assigned to Player A and the second to Player B.

of strategies), and in others the population coevolved by playing other individuals. The details of this work are reported in Axelrod (1986) and Forrest (1985). To summarize, we found that mutually cooperative strategies (receiving payoffs from the (3,3) box) did evolve from random starts and even from populations that were initialized with mutually defecting strategies. Furthermore, these cooperative strategies were stable—they could resist invasion by other noncooperative strategies.

This work showed that cooperative strategies could evolve under the genetic algorithm, both in the two-agent case described above and in an extension to the n-agent case, called Meta-Norms (Axelrod 1986). In this work, the genetic algorithm was viewed as a model of imitation; successful members of the population were imitated by less successful members, sometimes with random variation (mutations) or by combining two successful strategies (crossover).

GENETIC ALGORITHMS AND RICHARDSON-TYPE MODELS

The nonlinear dynamical systems approach of Mayer-Kress is compared with the genetic algorithms approach of Axelrod and Forrest in Table 13.1.

This comparison suggests many ways in which the two approaches can

P.D./Norms Games	Richardson-type Models
Goal: maximize payoff	Goal: maximize stability
Individual agents	Countries
Score	Defense Efforts
Strategy	Parameters
Moves (cooperate, defect)	Update rule (difference equations)
Genetic algorithm	Change parameters

Table 13.1: Comparison of genetic algorithms and Richardson-type models.

be combined to create systems with the advantages of both. The Richardson models provide a way to study the long-term implications of various policies (i.e., which policies are stabilizing and which are destabilizing), while the genetic algorithm provides a way to model the evolution of policies over time. The genetic algorithm also provides an effective way to search for parameter values that have certain properties (stability, minimal expenditures, etc.). In our initial implementation we have constructed a very simple hybrid system, substituting the last row of the P. D. column for the last row of the Richardson column. Thus, the genetic algorithm is being used to evolve the parameters of the three-agent Richardson equation, replacing the earlier method of setting parameters by hand.

This simple combination of the two approaches illustrates both aspects of the genetic algorithm in which we are interested: (1) the optimizer that can search parameter spaces efficiently, and (2) a model of negotiation. In the second aspect the genetic algorithm is modeling a negotiation process in the sense that it is searching for a set of parameters (policies) that satisfy some global criteria (stability, reduction of expenditures, etc.) and by which all countries might agree to abide. It has even been suggested that the crossover operator could be construed as a form of compromise between two competing policies. However, in contrast with the Prisoner's Dilemma example, our model does not allow each country to optimize its own parameter settings to maximize individual advantage (see the section "Future Directions").

The Representation

This section describes how the genetic algorithm was used to evolve parameters in the Richardson model. The two aspects of the specification are the fitness function F and the representation of parameters on a bit string.

The overall goal was to quickly identify interesting parameter sets and then study them in detail using other techniques. As a result, the fitness function was chosen to provide a crude estimate of the appropriateness of the parameters. This allows us to trade off the cost of evaluating the fitness against the number of points we sample. One question we were interested in answering was how good an estimate the fitness function F provided (see the subsection "Implications of the Results for the Model"). For the initial implementation, we have concentrated on parameter settings that result in a stable balance of power. As an estimate, we iterate the model for n time steps (typically $n = 20$) and compare the spending of X (the dominant country) with that of Y and Z (the allied countries):

$$F = |x_n - (y_n + z_n)|.$$

The simulation is designed to be run with either twelve or fifteen parameters. The additional three parameters are the initial conditions x_0, y_0, and z_0. When the parameters are included in the simulation, the initial conditions are subject to adaptation and are under the control of the genetic algorithm. When the parameters are not included, they are set to some fixed initial value, typically, $x_0 = y_0 = z_0 = 0.0$. Other possible fitness functions might minimize total expenditures, establish one country as the dominant power, or combine some of these factors. In the model there are twelve independent parameters (fifteen if the initial conditions are included). These real-valued parameters are defined to be in the range $[0, 1]$, so it is straightforward to discretize them into bins. This results in one binary-coded integer for each parameter. The integer designates in which bin the parameter lies. The size of the bins is determined by the number of bits used in the discretization. We have tried several different values and are currently using eight bits. This results in $2^8 = 256$ different bins for each parameter. The binary-coded integers (bins) are then gray-coded and concatenated in the following order:

$$k_{11}k_{22}k_{33}k_{23}k_{12}k_{13}x_sy_sz_sx_my_mz_mx_0y_0z_0$$

This results in a bit string of length 120 and a search space with 2^{120} different values.[1]

---

[1] It is important to be careful when estimating the size of the search space. Increasing the number of bins beyond the inherent sensitivity of the problem can inflate the size of the search space. On the other hand, it is important to make the number of bins large enough to reflect all the relevant different behaviors. For our problem, we determined the number of bins from modeling constraints and the intrinsic noise tolerance of the model. Economic growth rates, defense budgets, etc. are typically estimated to an accuracy of 0.1 percent. This suggests that the bin size should be roughly $1/1000$, which translates into 1000 bins for our model (since all parameters are in $[0, 1]$). However, past

The results reported in this chapter were generated using the public domain GENESIS package (Grefenstette 1984) on a Sun SPARC-1 running Sun OS 4.03. Typical runs use populations of Size 200-300, runs of about one hundred generations, crossover rates of 0.6,[2] and mutation rates of 0.00017. (Each bit has a probability 0.00017 of being mutated in any generation.)

RESULTS

The genetic algorithm finds balance-of-power solutions that were not known to exist previously. Manually setting the parameters had not suggested that such points existed. Figure 13.3 illustrates how the model behaves as the parameter set is perturbed from nearly perfect fitness. Each subfigure shows a different parameter setting: (a) k_{11}, k_{22}, k_{33}, k_{23}, k_{13}, k_{12}, x_s, y_s, z_s, x_m, y_m, z_m = 0.828, 0.473, 0.894, 0.7, 0.410, 0.859, 0.45, 0.445, 0.328, 0.683, 0.504, 0.180; (b) same as (a) except $k_{23} = 0.5$ and $x_s = 0.5$; (c) same as (a) except $k_{23} = 0.3$ and $x_s = 0.45$; (d) same as (a) except $k_{23} = 0.08$ and $x_s = 0.45$. Within a subfigure, the parameters are held constant, but the initial conditions are varied. The two-dimensional projections of the three-dimensional initial-condition space show initial conditions for Nations X and Y. The colors show which country is dominant after 111 time steps or if there is an unbounded arms race: black indicates that Country X is dominant, grey indicates that Country Y is dominant, and white indicates an unlimited arms race.

In Figure 13.3a fitness was 0.0 within the precision of our coding. For almost all initial conditions in the unit cube, the time evolution tends to the same arms control fixed point with perfect balance of power. In Figure 13.3b-d as the fitness moves away from 0.0, the surface becomes more irregular. An irregular surface means that it is difficult to predict the result of a particular parameter setting, since small changes to initial conditions can lead to vastly different results. These effects are even more pronounced when the three-dimensional surface is shown (initial conditions for all three countries). Figure 13.3a shows example results found by the genetic algorithm; the comparison with Figure 13.3d is striking. Figure 13.3a

experience indicates that noise has an effect on the model at a coarser level (closer to five bits). We have chosen eight bits of precision (256 bins) as a compromise between these two considerations.

[2]GENESIS uses two-point crossover (two crossover locations are chosen for each "crossover"). The crossover rate specifies the probability that an individual will undergo a two-point crossover per generation.

Parameter	Sample 1	Sample 2	Sample 3	Sample 4	Sample 5
K11	0.99	0.03	0.23	0.77	0.29
K22	0.37	0.78	0.48	0.69	0.95
K33	0.85	0.02	0.72	0.78	0.50
K23	0.83	0.94	0.87	0.58	0.04
K12	0.56	0.84	0.53	0.56	0.80
K13	0.28	0.31	0.02	0.91	0.89
XS	0.98	0.56	0.25	0.64	0.98
YS	0.71	0.59	0.39	0.71	0.74
ZS	0.84	0.36	0.65	0.84	0.05
XM	0.68	0.55	0.61	0.70	0.68
YM	0.24	0.41	0.59	0.24	0.24
ZM	0.44	0.96	0.02	0.44	0.44
X0	0.39	0.99	0.21	0.27	0.66
Y0	0.79	0.00	0.04	0.69	0.45
Z0	0.66	0.15	0.50	0.27	0.14

Table 13.2: Example Points Found by the Genetic Algorithm

is representative of the many points with Fitness 0.0 that we studied. Figures 13.3b through 13.3d show how the surface becomes progressively more fragmented as a point with nearly perfect fitness is changed by adjusting one parameter. These surfaces were generated without noise. We expect that they would be even more fragmented with noise.

These results are nontrivial in that the genetic algorithm found many different solutions and most of the solutions were not degenerate. (That is, parameters were not set to 0.0 or 1.0). For example, Table 13.2 shows five points, each of which has perfect fitness, that were found by one run of the genetic algorithm. Taken together, they represent a wide diversity of solutions.

In the next two subsections, we explore the significance of these results, first from the modeling perspective and second as an optimization problem.

Implications of the Results for the Model

Discovering that nontrivial balance-of-power points exist raises several questions: (1) Why is balance of power a good predictor of overall stability in the model? (2) What happens to the model as we perturb the balance-of-power solution? (3) How do the various parameters contribute to the balance-of-power solution? and (4) What is the relation between balance of power and fixed points in the model?

One surprising result is that the fitness function is an excellent predictor

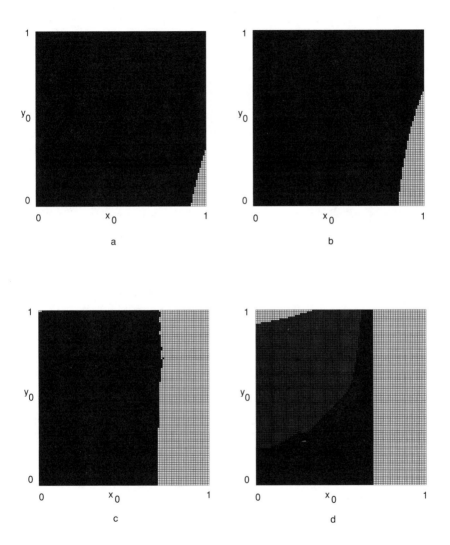

Figure 13.3: Basin boundary for balance-of-power parameter setting found by the genetic algorithm.

of stability in the model. It is surprising because the fitness function considers only one set of initial conditions and iterates the equations a small number of times (20). The low number of iterates can be explained by studying how fast the model converges (either to a fixed point or to infinity). Several tests were performed for various parameter configurations, and they all indicate fast convergence (within 20 time steps) unless the parameters are near a basin boundary or bifurcation. In regions near basin boundaries and bifurcations, the transient times to asymptotic behavior typically increase. Thus, we expect that in these regions our estimate will assign high fitness (a bad value) to some points that would eventually converge to stability if the system were iterated longer. This is advantageous from the application point of view since the points that our fitness function misses are exactly those that lie in an unstable region of the space. This is because the points that lie near basin boundaries or bifurcations are those for which the perfect balance of power is increasingly sensitive to small-scale perturbations.

Although the low number of iterates is understandable, it is quite surprising that an arbitrary set of initial conditions predicts well the stability of the model for all initial conditions. This result suggests that if the system is balanced ($Fitness \approx 0.0$), then the initial conditions do not matter, and if the system is not balanced, then its behavior is highly dependent on the initial conditions.

The genetic algorithm is searching for balance-of-power points. Yet, previous analysis of the system has focused on the existence of stable fixed points. We are currently studying the system analytically to determine how these two relate: How many stable fixed points does the model have? How do the fixed points change with balance of power?

Studying hyperplanes of parameter space shows that the parameter space contains regular subregions with perfect fitness. This means once we have found one solution with the genetic algorithm, it is easy to find other solutions in a neighborhood due to the smoothness of the system. Figure 13.4 illustrates the structure of the fitness function in a two-dimensional hyperplane through a single point. We started with the point

$$k_{11}k_{22}k_{33}k_{23}k_{12}k_{13}x_sy_sz_sx_my_mz_mx_0y_0z_0 =$$

$$0.8\ 0.5\ 0.9\ 0.5\ 0.4\ 0.9\ 0.6\ 0.4\ 0.3\ 0.7\ 0.5\ 0.2\ 0.4\ 0.4\ 0.3$$

This point is one of those found by the genetic algorithm (see Section 13 for the approximate values). We varied x_s and k_{23}. The fitness function evaluated the point as having fitness zero (within our tolerance). For the figure, nine parameters and the initial conditions are fixed at the genetic

algorithm values, parameters x_s (x-axis) and k_{23} (y-axis) are varied in
the interval [0,1], and k_{11} is fixed at a different value for each subfigure:
(a) $k_{11} = 0.1$, (b) $k_{11} = 0.2$, $k_{11} = 0.6$, $k_{11} = 0.8$. The axes represent
the unit interval [0,1] (evaluated at resolution $\delta x = \delta y = 0.01$). The
fitness of each point in the xy plane is shown; grey values go from black
$= 0$ (lower left corner of the grey chart) to white $=$ maximum fitness
(upper right corner). Note the continuous change of the area with 0.0
fitness in Figures (a) through (d), demonstrating the structure of the three-
dimensional projection of the balance-of-power surface.

The contour plot shows a blank region for which the fitness function is
almost equal to zero in the upper right part of the diagram. For small values
of x_s and k_{23} we observe a very steep increase in the fitness function. As
we move this slice in the direction of increasing values of k_{11}, these regions
deform in a very complex manner indicating the overall structure of the
fitness region in the fifteen-dimensional parameter space.

To summarize, the three-nation Richardson model is a high-dimensional
nonlinear system. The model is difficult to study analytically. The genetic algorithm has found structure in the parameters defining the model
that was not previously known, and it is providing insight about where
to look for analytical solutions. Specifically, the contour plots show that
balance-of-power points lie in hyperplanes through parameter space. This
is intriguing because according to the Schema Theorem the genetic algorithm searches the space of hyperplanes for those with good fitness. Since
the genetic algorithm found good solutions so easily, it is tempting to attribute this success to the relevance of hyperplanes to the solution.

Is the Problem GA-Hard?

From the viewpoint of optimization, an important question is how hard is
the problem? We consider a problem hard if it cannot be solved by standard
linear search techniques. In this subsection we investigate this question
and conclude that the problem is not particularly difficult from the genetic
algorithm perspective but that the solutions found by the genetic algorithm
are qualitatively different from those found by a linear search technique.
Stochastic iterated hill climbing was used as the baseline of comparison.

First, we studied the distribution of fitnesses for a random population.
Figure 13.5 shows this distribution. The figure shows that about 2.5 percent
of the random population is close to Fitness 0.0. This indicates that the
problem is not especially hard from an optimization point of view, since acceptable solutions could be found fairly quickly by randomly searching the
parameter space. This is quite surprising for several reasons: (1) the search

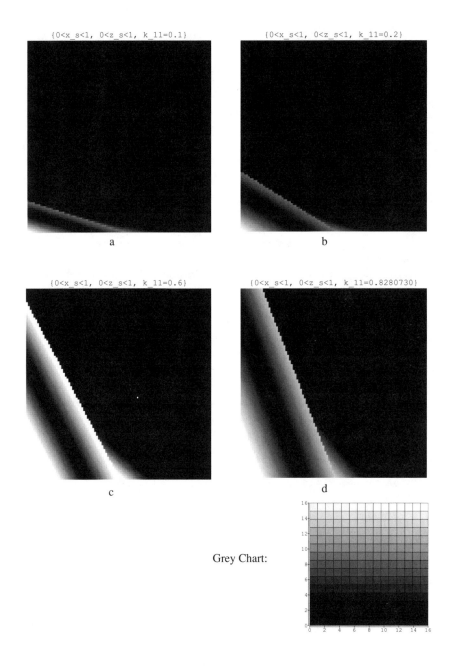

Figure 13.4: Contour plots.

Parameter	Sample 1	Sample 2	Sample 3	Sample 4	Sample 5
K11	0.13	0.11	0.22	0.16	0.31
K22	0.00	1.00	0.62	0.20	1.00
K33	0.00	1.00	1.00	0.84	1.00
K23	0.50	0.06	0.39	0.44	0.16
K12	0.50	0.61	0.53	0.50	1.00
K13	0.07	0.49	1.00	0.36	1.00
XS	0.63	0.82	0.85	0.02	0.30
YS	0.71	0.16	0.26	1.00	1.00
ZS	0.40	1.00	0.22	1.00	1.00
XM	0.71	0.75	0.89	0.02	1.00
YM	0.40	0.63	0.58	0.44	0.78
ZM	0.34	0.43	0.31	0.42	0.16
X0	0.03	0.00	0.00	0.00	0.50
Y0	0.00	0.00	0.00	0.00	0.00
Z0	0.47	0.49	0.31	0.71	0.56

Table 13.3: Example Points Found by Hill Climbing

space is enormous (2^{120}), (2) hand tuning had never revealed a balance-of-power solution, and (3) the problem is known to be nonlinear. Thus, the genetic algorithm is finding some structure in the problem that has not been revealed by other methods. We are currently trying to understand that structure.

We ran a discrete hill-climbing algorithm on the problem, and it was able to find solutions with perfect fitness within 1000 trials (function evaluations). However, the points that it found were quite different from those found by the genetic algorithm. Table 13.3 shows five example points found by the hill-climbing algorithm. Since all the parameters are normalized to the range [0,1], the large number of 1s and 0s in Figure 13.3 indicates solutions that are partially degenerate. Thus, both the genetic and hill-climbing algorithms can find points with perfect fitness. However, the points found by the genetic algorithm are qualitatively different from those found through hill climbing.

FUTURE DIRECTIONS

We hope to integrate the two approaches described in Table 13.1 more completely in the future, so that we can address several of the limitations in the current model. Specifically, the current system does not allow individual countries to optimize their own policies; rather, "God" solves the

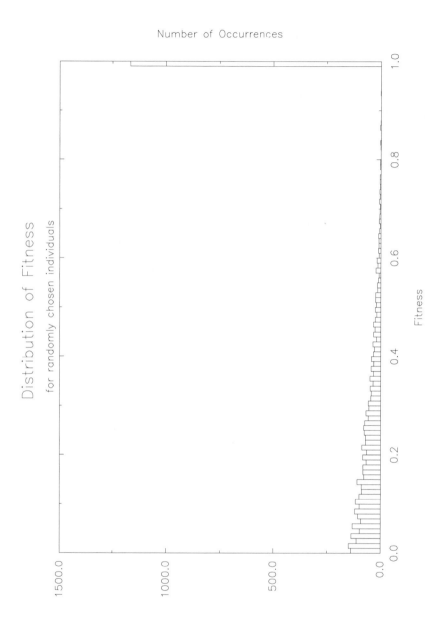

Figure 13.5: Histogram of frequency of fitness for population of random individuals. 5000 sample points in parameter space were generated randomly. The distribution of their fitnesses is shown as a histogram. All points with $fitness > 1.0$ are plotted at the point 1.0.

system of equations, resulting in a global solution such as one that might result from negotiated treaties. We envision a model that is more along the lines of the Prisoner's Dilemma work, in which completely self-interested, autonomous agents learn to cooperate over time. A second limitation is that alliances are formed in a fixed way—the two weakest countries are always allied against the strongest. We would prefer a model in which alliance formation is under the control of each individual country. Finally, the time-scale on which policies change needs to be on a par with changes in expenditures.

To address these limitations, we propose a model in which each country is represented by its own population of bit strings. Thus, there would be three different populations, each evolving independently under the genetic algorithm. In this way, each country could optimize its own behavior. Each country would make decisions about how to allocate its own resources (among military expenditures, domestic spending, and capital reinvestment) and with whom it allied itself. Alliance formation would become a kind of bidding game in which alliances were offered and accepted. Each country's fitness would depend on a combination of its internal stability and global interactions with the other countries.

Recent developments in Eastern Europe have demonstrated how structures that were rigid for decades can suddenly destabilize, leading to crises and qualitative changes. The end of a simple world with two major adversaries might indicate the beginning of a "messy future" with many relevant players. International security may also be increasingly influenced by other factors such as global climate changes. A large scientific community is currently building detailed models to describe this complex global system. As our simple model illustrates, there is a wide variety of possible consequences for political decisions in such a system. Computational methods, such as the genetic algorithm discussed in this chapter, will be increasingly important in suggesting possible alternatives. Several groups and institutions, including the Joint Chiefs of Staff and the Rand Corporation, have expressed interest in using global models similar to those discussed here, for a quick global analysis of complex situations (e.g., in arms-control negotiations). In these situations, policy decisions are often made in a very short time, precluding the use of extensive and detailed simulations.

CONCLUSIONS

We have taken a simple model and used it to illustrate how genetic algorithms can augment the standard computational and analytical techniques for studying nonlinear dynamical systems. The genetic algorithm was used successfully to optimize parameters in a nonlinear dynamical system. We expect that the methods described here will prove useful to a wide range of other nonlinear dynamics parameter optimization problems. We are currently trying the method on other nonlinear problems. Our project is the first one that we know of that attempts to use the genetic algorithm to find "good" parameter values for this class of systems.

This problem, however, does not appear to be a good example of a GA-hard problem. That is, the problem we have solved was not especially difficult for the genetic algorithm and probably could have been solved using other methods. For this application, modeling considerations are relevant, and the genetic algorithm provides a reasonable model for negotiation.

Even though the model is too simple to make reliable predictions about real arms races, it is useful because it presents a global view. More detailed models are not necessarily more reliable, because of intrinsic chaos.

ACKNOWLEDGMENTS

The authors gratefully acknowledge the help of Judith Challenger and Toshi Ohsumi who developed and ran the programs that produced the visualizations of our results. The work has benefited from discussions with John Holland. Several researchers from the Center for Nonlinear Studies have contributed suggestions about how to study the system analytically. These include Daniel David, Greg Forest, Daryl Holm, and Jia Li. We would also like to thank David Davis for his encouragement, patience, and many helpful suggestions.

REFERENCES

Axelrod, Robert (1984). *The Evolution of Cooperation.* New York: Basic Books.

Axelrod, Robert (1986). An evolutionary approach to norms. *The American Political Science Review* 80, December.

Axelrod, Robert (1987). The evolution of strategies in the iterated prisoner's dilemma. In L.Davis (ed.), *Genetic Algorithms and Simulated Annealing.* London: Pitman Publishing.

Challinger, Judith (1990). Interactive graphical exploration of the 3-nation Richardson model. University of California at Santa Cruz CIS Department preprint.

Eckmann, J. P. and David Ruelle (1983). Ergodic theory of chaos and strange attractors. *Review of Modern Physics* 57, p. 617.

Farmer, J. Doyne (1989). Personal communication.

Forrest, Stephanie (1985). Documentation for prisoner's dilemma and norms programs. The University of Michigan.

Grefenstette, John J. (1984). GENESIS: a system for using genetic search procedures. *Proceedings of a Conference on Intelligent Systems and Machines.* Rochester, Mi. pp. 161-5.

Grossmann, Siegfried and Gottfried Mayer-Kress (1989). Chaos in the international arms race. *Nature* 337, February, pp. 701-704.

Mayer-Kress, Gottfried (1989). A nonlinear dynamical systems approach to international security. Technical Report LA-UR-89-1355, Los Alamos National Laboratory.

Saperstein, Alvin M. and Gottfried Mayer-Kress (1988). A nonlinear dynamical model of the impact of SDI on the arms race. *Journal of Conflict Resolution* 32, 4. December, pp. 636-670.

Richardson, L. F. (1960). Arms and insecurity. *Boxwood.*

Richardson, L. F. (1960). Statistics of deadly quarrels. *Boxwood.*

Tucker, A. W. (1950). A two-person dilemma. Stanford University Tech Report.

14

Strategy Acquisition with Genetic Algorithms

John J. Grefenstette

INTRODUCTION

The growing interest in genetic algorithms can largely be attributed to the generality of the approach. Genetic algorithms can be used for both numerical parameter optimization (Fitzpatrick and Grefenstette 1988; Grefenstette 1986) and combinatorial search (Grefenstette 1987). This chapter shows an application to a rather different sort of problem: the optimization of policies for sequential decision tasks. In this approach, each policy, or *strategy*, is represented as a set of condition/action rules. Each proposed strategy is evaluated on a simulation model of the sequential decision task, and a genetic algorithm is used to search for high-performance strategies. The approach has been implemented in a system called SAMUEL. (SAMUEL stands for Strategy Acquisition Method Using Empirical Learning. The name also honors Art Samuel, one of the pioneers in machine learning.)

The chapter is organized as follows. First comes a discussion of the general problem area for which SAMUEL was designed—sequential decision tasks. Next, the operation of SAMUEL is outlined, focusing on a particular task—the *evasive maneuvers* problem. The next section describes the customizations of the basic genetic algorithm that have been adopted in SAMUEL. The final section discusses the applications of the method. This brief chapter should give the reader an idea of how genetic algorithm can be used to optimize strategies for this broad class of problems. For a more detailed discussion, see Grefenstette, Ramsey, and Schultz (1990).

SEQUENTIAL DECISION TASKS

Sequential decision tasks may be characterized by the following general scenario. A decision-making agent interacts with a discrete-time dynamical system in an iterative fashion. At the beginning of each time step, the agent observes a representation of the current state and selects one of a finite set of actions, based on the agent's decision rules. As a result, the dynamical system enters a new state and returns a (possibly null) payoff. This cycle repeats indefinitely. The objective is to find a set of decision rules that maximizes the expected asymptotic payoff rate. For a more precise mathematical formulation of sequential decision problems, see Barto, Sutton, and Watkins (1989). Several sequential decision tasks have been investigated in the machine learning literature, including pole balancing (Selfridge, Sutton, and Barto 1985), gas pipeline control (Goldberg 1983), and the animat problem (Wilson 1985, 1987). For many interesting problems, including those considered here, payoff is delayed in the sense that non-null payoff occurs only at the end of an episode that may span several decision steps. In fact, the paradigm is quite broad since it includes any problem-solving task by defining the payoff to be positive for any goal state and null for non-goal states (Barto, Sutton, and Watkins 1989).

Sequential decision problems include many important practical problems, and much work has been devoted to their solution. The field of adaptive control theory has developed sophisticated techniques for sequential decision problems for which sufficient knowledge of the dynamical system is available in the form of a tractable mathematical model. For problems lacking a complete mathematical model of the dynamical system, dynamic programming methods can produce optimal decision rules, as long as the number of states is fairly small. The Temporal Difference (TD) method (Sutton 1988) addresses learning control rules through incremental experience. Like dynamic programming, the TD method requires sufficient memory (perhaps distributed among the units of a neural net) to store information about the individual states of the dynamical system (Barto,

Sutton, and Watkins 1989). For problems that have enormous state spaces, these memory constraints may require that the state space be partitioned in order to apply TD methods. The approach described here focuses on rules rather than states. Since the number of rules explicitly considered in SAMUEL is constrained to be much less than the number of states, this approach seems more likely to be applicable to realistic problems. Since the generality of the rules (i.e., the number of states matched) is modified dynamically, SAMUEL is also less dependent on an *a priori* partition of the state space.

The system described in this chapter was designed to explore the application of genetic algorithms to control problems arising in multi-agent, tactical domains. In such domains, a complete mathematical analysis is usually impossible, because of the complexity of the multi-agent interactions and the inherent uncertainty about the future actions of other agents. One manual approach to such problems is to develop a policy expressed as a set of tactical rules, or *strategy*, that specify appropriate responses to any given situation. The behavior of a strategy can be monitored in a simulation to discover any weaknesses or inadequacies. Knowledge engineers can use this information to modify the rules, which can then be re-evaluated in the simulation. Such a generate-and-test cycle can be repeated until a satisfactory set of rules is found. In many applications, off-line testing of the system is the only realistic alternative for evaluating the performance of hypothetical strategies, since testing strategies on the "live" system is too difficult, too costly, or too dangerous. The current system was designed with off-line learning in mind. The goal is to reduce or eliminate the manual effort involved in the generate-and-test cycle in evolving high-performance strategies. In the remainder of this chapter, we describe our approach to the design of systems that can automatically assess and modify their heuristic rules for sequential decision tasks, given a simulation model of the task environment.

SAMUEL

SAMUEL is a system that uses a genetic algorithm (Holland 1975) to optimize strategies for sequential decision tasks. The design of SAMUEL reflects certain assumptions about the task environment. It is assumed that a fixed set of sensors is available to provide information concerning the current state and that a fixed set of control variables may be set by the decision making agent. A *strategy* in SAMUEL is a set of condition/action rules of

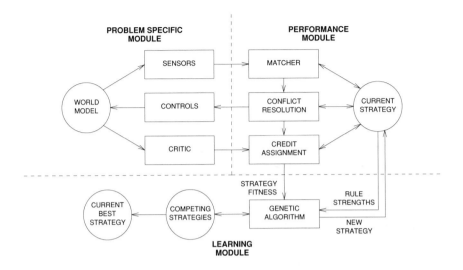

Figure 14.1: SAMUEL: A system for learning strategies.

the form

if   (and $c_1 \cdots c_n$)

then   (and $a_1 \cdots a_m$)

where each c_i is a condition on one of the sensors and each action a_j specifies a setting for one of the control variables. The system architecture is shown in Figure 14.1.

SAMUEL consists of three major components: a problem-specific module, a performance module, and a learning module. The problem-specific module consists of the task environment simulation, or world model, and its interfaces. The performance module is called CPS (Competitive Production System), a production system that interacts with the world model by reading sensors, setting control variables, and obtaining payoff from a critic. In addition to matching, CPS implements conflict resolution as a competition among rules based on rule strength and performs credit assignment based on payoff (Grefenstette 1988). The learning module uses a genetic algorithm to search for high performance strategies. Each strategy is evaluated on a number of tasks in the world model. As a result of these evaluations, genetic operators, such as CROSSOVER and MUTATION, produce plausible new strategies from high-performance parents.

Smith's LS-1 (Smith 1980) was the first system in which each structure in the population of a genetic algorithm was interpreted as a set of rules. This approach was extended by Schaffer's LS-2 (Schaffer 1984). SAMUEL not only builds on these earlier systems but it also draws on some ideas from classifier systems (Holland 1986). However, in a departure from earlier genetic learning systems, SAMUEL learns strategies consisting of rules expressed in a high-level rule language. An example of the rule representation is shown below.

We initially tested SAMUEL in the context of a sequential decision problem called Evasive Maneuvers (EM). In EM, there are two objects of interest: a plane and a missile. The decision maker controls the turning rate of the plane to avoid being hit by the approaching missile. The missile can track the motion of the plane and steer toward the plane's anticipated position. Six sensors give information about the current state: the current turning rate of the plane, a clock, the missile range, the missile bearing, the missile heading, and the missile speed. There is a single control variable, the turning rate of the plane. The process is divided into *episodes* that begin with the missile approaching the plane from a randomly chosen direction. The missile initially travels at a far greater speed but is less maneuverable than the plane (i.e., the missile has a greater turning radius than the plane) and gradually loses energy as it maneuvers. The episode ends when either the plane is hit or the missile's speed drops below a threshold and it loses maneuverability. This requires between 2 and 20 decision steps, depending on how many turns the missile performs while tracking the plane. At the end of each episode, the decision maker obtains a payoff defined by the formula:

$$payoff = 1000 \text{ if plane escapes missile}$$
$$= 10t \text{ if plane is hit at time } t$$

EM does not seem to be solvable by any simple, fixed strategy that ignores the position of the missile, such as making tight loops or flying straight ahead at full speed.

A strategy for EM consists of a set of decision rules, for example:

```
if   (and (last-turn 0 45)
          (time 4 14)
          (range 500 1400)
          (bearing 3 6)
          (heading 90 180)
          (speed 50 850))
```

```
then (and (turn 90))
strength 750
```

Each condition on the left-hand side of a rule specifies a range over the named sensor, and each action on the right-hand side specifies the value for the named control variable. The strength is an estimate of the rule's utility and is used for conflict resolution (Grefenstette 1988). The use of a high-level language for rules offers several advantages over low-level binary pattern languages typically adopted in genetic learning systems (Goldberg 1983; Smith 1980). First, it makes it easier to incorporate existing knowledge, whether acquired from experts or by symbolic learning programs. Second, it is easier to transfer the knowledge learned to human operators. Third, it makes it possible to combine empirical methods such as genetic algorithms with analytic methods that explain the success of the empirically derived rules.

CUSTOMIZATIONS OF THE GENETIC ALGORITHM

SAMUEL treats the learning process as a heuristic optimization problem, that is, as a search through a space of knowledge structures looking for structures that lead to high performance. A genetic algorithm is used to perform the associated search. Since genetic algorithms have been described earlier in this volume, we will restrict our discussion to the ways in which the standard genetic algorithm has been customized for this application.

Unlike the majority of genetic algorithms in Part I of this book, SAMUEL's population is completely renewed each iteration. This approach reflects SAMUEL's heritage as an outgrowth of the GENESIS software package. GENESIS implements what might be called a *generational* approach, whereas most of the OOGA modifications are adapted to the *steady-state* approach. The generational approach is outlined in Figure 14.2.

At iteration t, a new population $P(t)$ of structures is formed in two steps. First, structures in the current population are selected to be reproduced on the basis of their relative fitness. That is, high-performing structures may be chosen several times for replication, and poorly performing structures may not be chosen at all. Second, the selected structures are recombined using idealized genetic operators to form a new set of structures that are finally evaluated as solutions to the given problem.

procedure *Generational GA*
begin
    $t \leftarrow 0$;
    initialize $P(t)$;
    evaluate structures in $P(t)$;
    while termination condition not satisfied **do**
    begin
        $t \leftarrow t + 1$;
        select $P(t)$ from $P(t-1)$;
        recombine structures in $P(t)$;
        evaluate structures in $P(t)$
    end
end.

Figure 14.2: A genetic algorithm.

In contrast, the steady-state approach selects two parent structures, re-combines them, evaluates them, and adds them to the population, replacing some older structures. These two approaches represent different ways to simulate an essentially parallel algorithm on a sequential computer. The mathematical differences between the approaches depend on the details of the selection algorithms, but these are beyond the scope of this chapter. Which approach is to be preferred is largely a matter of taste—each offers distinct implementation advantages. In SAMUEL, each strategy is evaluated by invoking CPS, using the given strategy as rule memory. CPS executes a small number (e.g., 20) of episodes, each episode starting from a randomly selected initial state. The average payoff of these episodes is returned as the fitness for the strategy. Since the evaluation of a single strategy in-volves running a simulator of the task environment, this paradigm maps easily onto coarse-grained multiprocessors. We have installed SAMUEL on a 128-node BBN Butterfly at the Naval Research Laboratory. By execut-ing a copy of CPS and the world model at each node, we can evaluate the entire population simultaneously (see Figure 14.3). This approach makes the system applicable to simulations of fairly complex real-world problems.

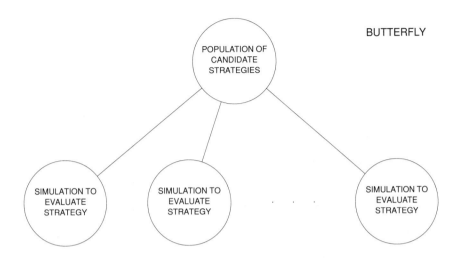

Figure 14.3: Exploiting parallelism in genetic learning.

INITIALIZATION

Research studies with genetic algorithms often assume a random initial population, in order to provide the maximum challenge for the learning algorithm. Some studies have investigated seeding the initial population with available knowledge (Grefenstette 1987). Although the rule language of SAMUEL facilitates the inclusion of heuristic rules, thus far we have concentrated on an approach that might be called *adaptive initialization*. Each strategy starts out as a set of completely general rules, but its rules are specialized according to its early experiences. That is, each rule in the initial population says:

    for any sensor readings, take action X

where X is one of the possible actions. A strategy consisting of only such rules executes essentially a random walk, since every rule matches on every cycle and all rules start with equal strength.

In order to introduce plausible new rules, a strategy modification operator called SPECIALIZE is applied after each evaluation of a strategy. SPECIALIZE is similar in spirit to Holland's *triggered operators* (Holland 1986). The trigger in this case is the conjunction of the following conditions:

1. There is room in the strategy for at least one more rule.

2. A maximally general rule is fired during a successful episode.

If these conditions hold, SPECIALIZE creates a new rule with the same right-hand side as the maximally general rule, but with a more specialized left-hand side. For each sensor, the condition in the new rule covers approximately half the legal range for that sensor, roughly centered on the sensor reading observed during the successful episode. For example, suppose the initial strategy contains the maximally general rule:

Rule 6: if ( ) then (and (turn 90))

Suppose further that the following step is recorded in the evaluation trace during the evaluation of this strategy:

Trace:

$\vdots$

t: sensors: ... (range 500) (speed 1000) ...

$\vdots$

Then SPECIALIZE would create the following new rule:

if   (and ... (range 250 750) (speed 500 1500) ...)
then   (and (turn 90))

assuming that *range* takes on values between 0 and 1000 and that *speed* takes on values between 0 and 2000. The resulting rule is given a high initial strength and is added to the strategy. The new rule is plausible, since its action is known to be successful in at least one situation that matches its left-hand side. Of course, the new rule is likely to need further modification and is subject to further competition with the other rules.

GENETIC OPERATORS: SELECTION, CROSSOVER, AND MUTATION

Strategies are selected for reproduction on the basis of their overall fitness scores returned by CPS. One interesting aspect of SAMUEL's selection algorithm concerns its approach to scaling (Grefenstette 1986). If the baseline against which the fitness of the individuals in a genetic algorithm are measured remains fixed, then as the overall performance rises within the population, the marginal differences between individuals decrease, and it becomes more difficult for the genetic algorithm to identify and reward the better individuals. The scaling problem is how to maintain selective pressure as the overall performance rises within the population. In SAMUEL, the fitness of each strategy is defined as the difference between the average payoff received by the strategy and a baseline performance measure that is adjusted to track the mean payoff received by the population, minus one standard deviation. The baseline is adjusted slowly to provide a moderately consistent measure of fitness. Strategies whose payoffs fall below the baseline are assigned a fitness measure of 0, resulting in no offspring. The effect is that, as the baseline rises, the genetic algorithm raises its standards. This mechanism appears to provide a reasonable way to maintain consistent selective pressure toward higher performance.

Selection alone merely produces clones of high-performance strategies. CROSSOVER works in concert with selection to create plausible new strategies. In SAMUEL, CROSSOVER treats rules as indivisible units. Since the rule ordering within a strategy is irrelevant, the process of recombination can be viewed as simply selecting rules from each parent to create an offspring strategy. Many genetic algorithms permit recombination within individual rules as a way of creating new rules (Schaffer 1984; Smith 1980). While such operators are easily defined for SAMUEL's rule language (Grefenstette 1989), we prefer to use CROSSOVER solely to explore the space of rule combinations, and leave rule modification to other operators (e.g., MUTATION and SPECIALIZE).

The CROSSOVER operator in SAMUEL attempts to cluster the rules before assigning them to offspring. For example, suppose that the traces of the most previous evaluations of the parent strategies are as follows ($R_{i,j}$ denotes the jth rule in strategy i):

Trace for parent #1:
Episode:

    $\vdots$

8. $R_{1,3} \to R_{1,1} \to R_{1,7} \to R_{1,5}$ Successful maneuver
9. $R_{1,2} \to R_{1,8} \to R_{1,4}$ Failure

    $\vdots$

Trace for parent #2:

    $\vdots$

4. $R_{2,7} \to R_{2,5}$ Failure
5. $R_{2,6} \to R_{2,2} \to R_{2,4}$ Successful maneuver

$$\vdots$$

Then one possible offspring would be:

$$\{R_{1,8}\,,\ldots,R_{1,3}\,,R_{1,1}\,,R_{1,7}\,,R_{1,5}\,,\ldots,R_{2,6}\,,R_{2,2}\,,R_{2,4}\,,\ldots,R_{2,7}\}$$

The idea is that rules that fire in sequence to achieve a successful maneuver should be treated as a group during recombination, in order to increase the likelihood that the offspring strategy will inherit some of the better behavior patterns of its parents. Rules that do not fire in successful episodes (e.g., $R_{1,8}$) are randomly assigned to one of the two offspring. Of course, the offspring may not behave identically to either one of its parents, since the probability that a given rule fires depends on the context provided by all the other rules in the strategy.

As explained previously, new rules are introduced through two operators, SPECIALIZE and MUTATION. MUTATION introduces new rules by making random changes to existing rules. For example, MUTATION might alter a condition within a rule from (range 500 1000) to (range 200 1000), or it might change the action from (turn 45) to (turn -90). In the current system, the new values produced by mutation are chosen randomly from the set of legal values for the given condition or action. The probability that a given value within a rule is mutated is less than 0.01 per generation. This level of mutation is consistent with previous studies (Grefenstette 1986). Many other forms of mutation have been incorporated into SAMUEL, including a version of Davis's (1989) CREEP operator that makes only small changes, for example, from (range 500 1000) to (range 400 1000). Future studies with SAMUEL will investigate such operators.

EXPRESSING CONSTRAINTS

Constraints in the rules of SAMUEL can be expressed in several ways. First, the user is required to define the legal range and the granularity for each sensor and control variable. For example, in the EM environment, the user might declare that heading has the range 0 to 360 degrees in 10 degree increments and that there are nine possible actions, consisting of turning from -180 to 180 degrees in 45 degree increments. The SPECIALIZE and MUTATION operators use these declarations to ensure that only semantically valid rules are generated. A second way to constrain the strategies learned is through the selective use of existing knowledge. The rule language of

SAMUEL was designed to facilitate the inclusion of existing knowledge in the same rule base as the learned rules. Including manually generated rules, with high initial strengths, can provide the learning system with a non-trivial starting point for its explorations. The definition of the learning operators provides another way to constrain the generation of rules. For example, the MUTATION operator can be constrained to allow only plausible ranges in the conditions of a rule. Finally, we can enforce constraints through the use of penalty functions (Richardson, Palmer, Liepins and Hilliard 1989). That is, strategies that violate some constraint can be given a very low fitness rating. The use of penalty functions should be avoided as much as possible, since it provides an indirect constraint on the genetic algorithm. It is preferable to constrain the representation and operators so that, for the most part, every strategy generated by the genetic algorithm satisfies at least some minimal requirements of plausibility for the task at hand.

In any case, it is important to design the evaluation function with care. Genetic algorithms are highly opportunistic optimizers and may produce surprising results if the fitness function rewards some behavior that the system designer does not want. There is an interesting story told by Mark Bramlette that illustrates this point: An engineer used a genetic algorithm to design a controller that could maximize the flight time for a proposed short-duration unmanned aircraft. Using a simulation of the proposed vehicle, the best manually designed controller yielded a flight time of about two minutes. A genetic algorithm was run on the problem overnight and produced a controller with a flight time of about three minutes. Encouraged, the engineer set the genetic algorithm to work over the weekend, and it produced a solution of about 10 minutes, which the design engineers thought was not possible. Upon further inspection, it was found that the "throttle" variable was not sufficiently constrained in the simulation model. In fact, the genetic algorithm had found a way to push the throttle negative after launch, thereby achieving a positive fuel gain! In this case, the model was easily fixed, and the resulting genetic algorithm once again converged on its original three-minute solution. The moral of the story is that a genetic algorithm will exploit opportunities in the model to maximize performance. Of course, this implies that the model should accurately reflect the constraints in the target environment to the greatest possible extent.

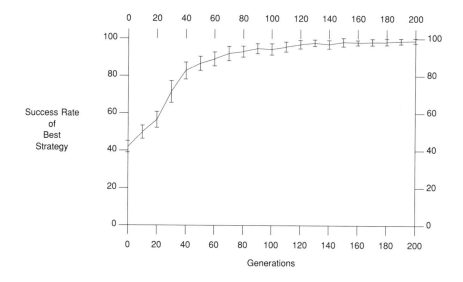

Figure 14.4: Typical learning curve with SAMUEL on EM problem.

APPLICATIONS

In recent experiments, SAMUEL has learned strategies for the Evasive Maneuvers problem that evade the missile over 99 per cent of the time. (A variety of simple strategies, such as making tight loops or making random turns, typically achieved success rates of about 40 per cent.) A typical learning curve is shown in Figure 14.4. Since SAMUEL employs probabilistic methods, the figure shows results averaged over 20 separate runs, with error bars indicating plus or minus one standard deviation.

Results reported elsewhere (Grefenstette, Ramsey, and Schultz 1990; Schultz, Ramsey, and Grefenstette 1990) indicate that this approach can be expected to scale up well to realistic tactical problems. Our initial series of studies have demonstrated that

- SAMUEL can learn strategies to evade missiles with a broad range of maneuverability characteristics.

- SAMUEL can learn high-performance strategies with noisy sensors.

- SAMUEL can learn robust strategies that perform well under a variety of initial conditions (e.g., missile starting speed and range).

- SAMUEL can effectively use available knowledge.

More extensive experiments with more complex versions of the EM task are currently underway. Of course, the learning techniques embodied in SAMUEL are in no way limited to the EM task. The application of these techniques to a number of simulations is under active evaluation at the Naval Research Laboratory.

As simulation technology improves, it will become possible to provide learning systems with high fidelity simulations of tasks whose complexity or uncertainty precludes the use of traditional knowledge engineering methods. The application of SAMUEL to simulations of military interest is expected to provide a number of measurable improvements in the design of tactical decision systems. In particular, we expect the results of this project to

1. Provide qualitatively better recommended tactical rules. By providing a set of rules that have been systematically developed and validated against a wide variety of simulated scenarios, we will reduce the effort required by human decision makers to tune fielded systems.

2. Reduce the time required to modify decision rules. By supplementing manual methods with automated techniques, we expect to reduce the time required to develop and validate strategies by an order of magnitude.

3. Expand the range of system utility. By inducing effective rules from an extensive range of simulated scenarios, we expect to improve the overall engagement effectiveness and resource utilization of the system by a significant factor.

4. Improve training opportunities. By using learning methods within training simulators, we expect to be able to provide more challenging and more realistic training environments for human operators.

Our initial studies with a simple tactical problem have shown that it is possible for learning systems based on genetic algorithms to effectively search a space of knowledge structures and discover sets of rules that provide high performance in a variety of target environments. Further developments along these lines can be expected to reduce the manual knowledge acquisition effort required to build systems with expert performance on complex sequential decision tasks.

REFERENCES

Barto, A. G., R. S. Sutton, and C. J. C. H. Watkins (1989). Learning and sequential decision making. COINS Technical Report, University of Massachusetts, Amherst.

Davis, L. (1989). Adapting operator probabilities in genetic algorithms. *Proceedings of the Third International Conference on Genetic Algorithms.* Fairfax, Va.: Morgan Kaufmann, pp. 61-69.

Fitzpatrick, M. J., and J. J. Grefenstette (1988). Genetic algorithms in noisy environments. *Machine Learning,* 3(2/3), pp. 101-120.

Goldberg, D. E. (1983). *Computer-Aided Gas Pipeline Operation Using Genetic Algorithms and Machine Learning.* Doctoral dissertation, Department of Civil Engineering, University of Michigan, Ann Arbor.

Grefenstette, J. J. (1986). Optimization of control parameters for genetic algorithms. *IEEE Transactions on Systems, Man, and Cybernetics, SMC-16(1),* pp. 122-128.

Grefenstette, J. J. (1987). Incorporating problem specific knowledge into genetic algorithms. In L. Davis (ed.), *Genetic Algorithms and Simulated Annealing.* London: Pitman Press.

Grefenstette, J. J. (1988). Credit assignment in rule discovery system based on genetic algorithms. *Machine Learning,* 3(2/3), pp. 225-245.

Grefenstette, J. J. (1989). Incremental learning of control strategies with genetic algorithms. *Proceedings of the Sixth International Workshop on Machine Learning.* Ithaca, NY: Morgan Kaufmann. pp. 340-344.

Grefenstette, J. J., Connie Loggia Ramsey, and A. C. Schultz (1990). Learning sequential decision rules using simulation models and competition. To appear in *Machine Learning.*

Holland, J. H. (1975). *Adaptation in Natural and Artificial Systems.* Ann Arbor, Mich.: University of Michigan Press.

Holland, J. H. (1986). Escaping brittleness: The possibilities of general-purpose learning algorithms applied to parallel rule-based systems. In R.S. Michalski, J. G. Carbonell, and T. M. Mitchell (eds.), *Machine Learning: An Artificial Intelligence Approach* (Vol. 2). Los Altos, Calif.: Morgan Kaufmann.

Richardson, J. T., M. R. Palmer, G. E. Liepins, and M. Hilliard. (1989). Some guidelines for genetic algorithms with penalty functions. *Proceedings of the Third International Conference on Genetic Algorithms.* Fairfax, Va.: Morgan Kaufmann, pp. 191-197.

Schaffer, J. D. (1984). *Some Experiments in Machine Learning Using Vector Evaluated Genetic Algorithms.* Doctoral dissertation, Department of Electrical and Biomedical Engineering, Vanderbilt University, Nashville.

Schultz, A. C., C. Loggia Ramsey, and J. J. Grefenstette (1990). Simulation-assisted learning by competition: Effects of noise differences between training model and target environment. *Proceedings of the Seventh International Conference on Machine Learning.* Austin, Tex.: Morgan Kaufmann, pp. 211-215.

Selfridge, O., R. S. Sutton, and A. G. Barto (1985). Training and tracking in robotics. *Proceedings of the Ninth International Conference on Artificial Intelligence.* Los Angeles, Calif.: Morgan Kaufmann.

Smith, S. F. (1980). *A Learning System Based on Genetic Adaptive Algorithms.* Doctoral dissertation, Department of Computer Science, University of Pittsburgh.

Sutton, R. S. (1988). Learning to predict by the method of temporal differences. *Machine Learning* 3, pp. 9-44.

Wilson, S. W. (1985). Knowledge growth in an artificial animal. *Proceedings of the International Conference on Genetic Algorithms and Their Applications,* Pittsburgh, Pa.: Lawrence Erlbaum Associates, pp. 16-23.

Wilson, S. W. (1987). Classifier systems and the animat problem. *Machine Learning* 2(3), pp. 199-228.

15

Genetic Synthesis of Neural Network Architecture

Steven A. Harp and Tariq Samad

INTRODUCTION

Technology periodically steals a leaf from nature's book. Both the genetic algorithm and the so–called connectionist or artificial neural network systems for computation are examples. Compared to the processing elements of conventional computers, the nervous systems of animals are composed of very low-speed devices (neurons), yet they possess unparalleled computing power (or perhaps we should say they possess very paralleled computing power). The strategy of deploying in parallel large numbers of very modest computing elements is one that artificial neural networks attempt to exploit. In most cases, the models used by computer scientists and psychologists are gross oversimplifications of physical neurons and their connections. Nonetheless, some of the properties illustrated in artificial neural network simulations reflect the properties of real neural systems. These include adaptability, generalization, and fault tolerance.

THE NETWORK DESIGN PROBLEM

Many artificial neural network models share another attribute with their biological counterparts: a lack of perspicuity. Their massive parallelism, nonlinearities, and adaptive characteristics all conspire to render analytic treatments of artificial neural networks fairly limited in scope. Advances in the theory are continually being made, but they are of little help in such important aspects of neural network applications as the choice of network structure and learning rule parameters. Our understanding of such aspects is primarily empirical; various rules of thumb are typically followed that are derived from experience in both practical and toy applications.

We address two major shortcomings of this heuristic approach here. First, the space of possible artificial neural network architectures is extremely large and, even with toy applications, most of it remains unexplored. It is simply impractical manually to evaluate a reasonable variety of architectures. Second, what constitutes a good architecture is intimately dependent on the application. Both the problem that is to be solved (e.g., handwritten character recognition) and the constraints on the neural network solution (e.g. fast learning and/or low connectivity and/or high accuracy) need to be considered, but at present we have no techniques or methods for doing so.

As a consequence of these shortcomings, some significant amount of manual trial-and-error experimentation is necessary before adequate performance is achieved, and no meaningful attempt is made to determine optimal architectures. Most applications adopt simple structures and conservative values of learning rule parameters. In particular, the structural aspects of artificial neural network design have been neglected. This is unfortunate because there is good reason to expect that the most effective artificial neural networks will exploit structural specializations of various sorts (Edelman 1988, Purves 1988).

RELEVANCE OF GENETIC APPROACHES

The problem of optimizing neural network structure for a given set of performance criteria is a complicated one. There are many variables, both discrete and continuous, and they interact in a complex manner. The evaluation of a given design is a noisy affair, since the efficacy of training depends on starting conditions that are typically random. In short, the problem is a logical application for the genetic algorithm.

We have been investigating the use of the genetic algorithm for designing application-specific neural networks (Harp et al 1989a). In our approach, realized in an experimental system dubbed NeuroGENESYS, the genetic algorithm is used to synthesize appropriate network structures and values for learning parameters. We focus more on the relationships between sets of units and bundles of connections than on determining each individual connection.

In fact, the genetic algorithm can be applied to the problem of neural network design in several ways. Montana and Davis (1989) and Whitley (1988, 1989) have explored the use of the genetic algorithm in training a network of known structure. Miller, Todd, and Hegde (1989) have used the genetic algorithm to apply constraints to the connection matrix of networks to be trained by backpropagation. Our approach is most similar to that of Schaffer, Caruana, and Eshelman (1989) who also use the genetic algorithm to discover the size, structure, and learning parameters of a network to be trained by a separate artificial neural network learning algorithm.

The appropriate way to apply the genetic algorithm to the study of neural networks (or the relevance of applying it at all!) depends on the questions we wish answered. We have two goals in our studies. The first is to develop a next-generation development tool for artificial neural networks; such a tool would allow a designer to describe the problem or class of problems to be solved and would automatically search for an optimal (by the user's criterion) network design. The second goal is to uncover further evidence to help build a theory of artificial neural network design. The genetic algorithm can help us answer questions about what kinds of structures will survive under a given set of conditions. It cannot tell us why these structures work or fail, but it can expose the "fossil record."

CONNECTIONIST ARCHITECTURES

Many paradigms for artificial neural networks are being explored, and a full survey of this burgeoning field is beyond the scope of any one book. We will sketch a few of the characteristics of the network paradigm relevant to this chapter. The interested reader is referred to Rumelhart and McClelland (1986) for more details. The typical network considered here is a directed acyclic graph of simple neurons or *units*. Each unit has a state, represented by a real number, a set of input connections, and a set of output connections to other units. The connections themselves have real-valued weights, w_{ij}.

A unit's state, o_i, is computed as a nonlinear function of the weighted sum of the states of units from which it receives inputs. The nonlinear activation function is a sigmoid, effectively endowing the unit with a threshold action; the position of the threshold is controlled with a bias or "threshold weight," θ_i. This is summarized in Equation 15.1.

$$s_i = \sum_{j=1}^{n} w_{ij} o_j + \theta_i \quad o_i = (1 + e^{-s_i})^{-1} \tag{15.1}$$

A subset of the units is designated as *input units*. These units have no input connections from other units; their states are fixed by the problem. Another subset of units is designated as *output units*; the states of these units are considered the result of the computation. Units that are neither input nor output are known as *hidden units*. A problem will specify a *training set* of associated pairs of vectors for the input units and output units.

The full specification of a network to solve a given problem involves enumerating all units, the connections between them, and setting the weights on those connections. The first two tasks are commonly solved in an ad hoc or heuristic manner, while the final task is usually accomplished with the aid of a learning algorithm, such as *backpropagation* (Werbos 1974, Rumelhart and McClelland 1986). Backpropagation is a supervised learning technique that performs gradient descent on a quadratic error measure to modify connection weights. A network begins with small random weights on its connections and is trained by comparing its response to each stimulus in the training set with the correct one and altering weights accordingly. The learning algorithm introduces new design variables, such as the parameters to control rate of descent.

THE BIG LOOP

The NeuroGENESYS system uses the genetic algorithm to search a space of possible neural network architectures. In most of our experiments, the system begins with a population of randomly generated networks. The structure of each network is described by a chromosome or genetic blueprint—a collection of genes that determine the anatomical properties of the network structure and the parameter values of the learning algorithm. We use backpropagation to train each of these networks to solve the problem, and we evaluate the fitness of each network in a population. We define fitness to be a combined measure of worth on the problem, which may take into account learning speed, accuracy, and cost factors such as the size and complexity

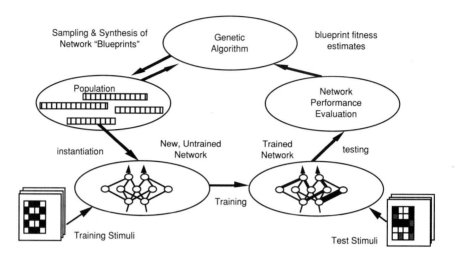

Figure 15.1: A population of "blueprints"—designs for different neural networks—is cyclically updated by the genetic algorithm based on their fitness scores. Fitness is estimated by instantiating each blueprint into an actual neural network, training, and then testing the network on the pattern recognition task—in this case identifying digits.

of the network. Network blueprints from a given generation beget offspring according to a reproductive plan that takes into consideration the relative fitness of individuals. In this respect, our application of the genetic algorithm is little different from any other function optimization application. A network spawned in this fashion will tend to contain some attributes from both of its parents. A new network may also be a mutant, differing in a few randomly selected genes from a parent. Novel features may arise in either case: through synergy between the attributes of parents or through serendipitous mutation. The basic cycle is illustrated in Figure 15.1.

This process of training individual networks, measuring their fitness, and applying genetic operators to produce a new population of networks is repeated over many generations. If all goes well, each generation will tend to contain more of the features that were found useful in the previous generation, and an improvement in overall performance can be realized over the previous generation.

Several interesting research issues are involved in using genetic algorithms for designing neural networks. These include the representation of the blueprint that specifies both the structure and the learning rule, the choice of the underlying space of network architectures to explore, adaptations of the genetic operators used to construct meaningful network structures, and the form of the evaluation function that determines the fitness of a network. The remainder of this chapter will discuss these issues and our approach to resolving them. We present some initial results from empirical studies of the problem.

REPRESENTATION

The development of a representation for neural network architecture is a major problem. Neuroscience is just beginning to understand how the plans for biological neural networks are represented (Edelman 1988). Unfortunately, biological neural networks are not yet understood well enough to provide clear guidelines for our synthetic problem. There are, of course, many different ways to parameterize network organization and operation. Network parameters might include the number of layers, the number of units in a layer, the number of feedback connections allowed, the degree of connectivity from one layer to another, the learning rate, and the error term utilized by the learning rule.

Ideally, a representation should be able to capture all potentially "interesting" networks, that is, those capable of doing useful work, while excluding flawed or meaningless network structures. It is obviously advantageous to define the smallest possible search space of network architectures that is sure to include the best solution to a given problem. An important implication of this goal in the context of the genetic algorithm is that the representation scheme should be closed under the genetic operators. In other words, the recombination or mutation of network blueprints should always yield new, meaningful network blueprints. There is a difficult trade-off between expressive power and the admission of flawed or uninteresting structures. We do not wish to limit the dimensionality of the network, nor do we want to allow it to grow out of control. We have formulated a network representation that, while unlikely to be ideal, appears to hold promise.

The gross anatomy of our prototype representation for neural network architectures is illustrated in Figure 15.2. Conceptually, all the network's parameters have been encoded in one long string of bits. The bit string is composed of one or more segments, each of which represents an area and its efferent connectivity, or projections. Areas need not be simply ordered, unlike most feedforward neural networks. Each segment is an area specification substring that consists of two parts: (1) an area parameter specification (APS) which is of fixed length and parameterizes the area in terms of its address, the number of units in it, how they are organized, and learning parameters associated with its threshold weights; and (2) one or more projection specification fields (PSFs), each of fixed length. Each field describes a projection from one area to another. As the number of layers is not fixed in our architecture (although it is bounded), the number of PSFs in an area segment is not fixed either. A projection is indicated by the address of the target area, the degree of connectivity, the dimension of the projection to the area, and so on.

The possibility that there may be any number of areas or projections motivates the use of markers with the bit string to designate the start and end of area segments and the start of PSFs. The markers enable a reader program to parse any well-formed string into a meaningful neural network architecture; they serve as landmarks but do not occupy any bits. The markers also allow a special genetic crossover operator to discover new networks without generating "nonsense strings."

Figure 15.2 shows how the APS and PSF are structured in our current representation. The portions of the bit string representing individual parameters are labeled boxes in the figure. They are substrings consisting of some fixed number of bits. Parameters described by an interval scale (e.g., 0, 1, 2, 3, 4) are rendered using Gray coding, thus allowing values that are close on the underlying scale to be close in the bit string representation (Bethke 1980, Caruana and Schaffer 1988). In our current system, most parameters are encoded with fields three bits wide, allowing eight possible values.

In the APS, each area has an identification number that serves as a name. The name need not be unique among the areas of a bit string. The input and output areas have the fixed identifiers 0 and 7. An area has a size and a spatial organization. The "total size" parameter determines how many units the area will have. It ranges from 0 to 7 and is interpreted as the logarithm (base 2) of the actual number of units; for example, if total size is 5, there are 32 units. The two "dimension share" parameters, which are also base-2 logarithms, impose a spatial organization on the units. The units of areas may have one- or two-dimensional rectilinear extent; in other words, the units may be arrayed in a line or rectangle. The network instantiation software attempts to find area dimensions whose product is equal to the desired total size while conforming closely to the indicated shares. The motivation for this organization comes from the sort of perceptual problems to which neural networks are apparently well suited. For example, an image processing problem may best be served

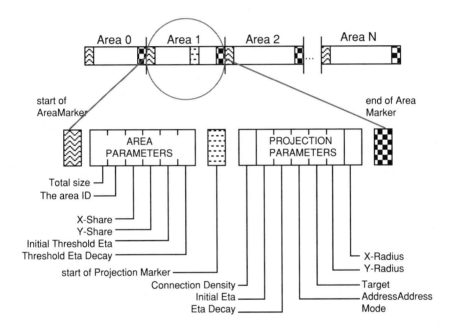

Figure 15.2: Overview of the prototype bit string representation of neural network architectures (top). The patterned bars are markers indicating the start and end of area segments. Detail shows the mapping of network parameters in the area specification fields and projection fields.

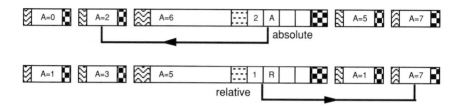

Figure 15.3: Examples of absolute and relative addressing in projections. Each of these individuals consists of five areas; the third area, shown in expanded form, sends a projection to another area.

by a square array, while an acoustic interpretation problem might call for vectors. The organization of the units in more conventional approaches is often left implicit; in our scheme dimensionality has definite implications for the architecture of projections. The PSFs in an area's segment of the bit string determine where the units in that area will make efferent connections and how. The representation scheme does not assume a simple pipeline architecture; different units within an area can be connected to different portions of a target area, and even to different target areas.

Each PSF indicates the identity of the target area. Our representation allows two ways to specify the target, distinguished by the value of a binary addressing mode parameter in each PSF. In the Absolute mode, the PSF's address parameter is taken to be the ID number of the target area. The Relative mode indicates that the address bits hold the position of the target area in the bit string relative to the *current area*. A relative address of zero refers to the area immediately following the one containing the projection; a relative address of n refers to the nth area beyond this, if it exists. Relative addresses indicating areas beyond the end of the blueprint are taken to refer to the final area of the blueprint—the output area. See Figure 15.3 for examples of absolute and relative addressing.

The different addressing schemes allow relationships between areas to develop, and to be sustained and generalized across generations through the genetic algorithm's reproductive plan. Specifically, the addressing schemes are designed to help allow these relationships to survive the crossover operator—either intact or with potentially useful modifications. Absolute addressing allows a projection to indicate a target no matter where that target winds up in the chromosome of a new individual. Relative addressing helps areas that are close in the bit string to maintain projections, even if their IDs change.

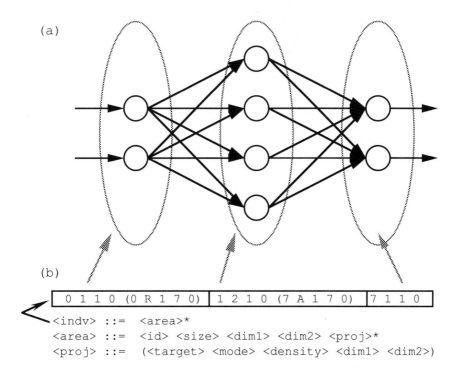

(a)

(b)

| 0 1 1 0 (0 R 1 7 0) | 1 2 1 0 (7 A 1 7 0) | 7 1 1 0 |

```
<indv>  ::=  <area>*
<area>  ::=  <id> <size> <dim1> <dim2> <proj>*
<proj>  ::=  (<target> <mode> <density> <dim1> <dim2>)
```

Figure 15.4: An example of a blueprint for a network with a single hidden area (a) and the structure of the network generated (b).

The dimension radii parameters (also base-2 logarithms) allow units in an area to project only to a localized group of units in the target area. This feature allows the target units to have localized receptive fields, which are both common in biological neural networks and highly desirable from a hardware implementation perspective. Even within receptive fields, projections between one area and another do not necessarily imply full factorial connectivity. The connection density parameter for the projection may stipulate one of eight degrees of connectivity between 30 percent and 100 percent.

Projections subsume one or more weight matrices; the weights are adjusted by the backpropagation learning rule during the training of the network. Parameters are included in the PSF to control the application of the learning rule for the projection. The eta parameter controls the learning rate in backpropagation and may take on one of eight values between 0.1 and 12.8. Eta need not remain constant throughout training; a separate eta-slope parameter controls the rate of exponential decay for eta as a function of the training epoch. Separate eta and eta decay parameters are included (in the ASF) for threshold weights. Figure 15.4 illustrates the structural parameters of our blueprint representation with a simple three-area network.

GENETIC OPERATORS

The reproductive plan used in our experiments is quite conventional; in each generation all individuals in the population are evaluated, and the population is subsequently sampled according to fitness. The sampling algorithm is based on the stochastic universal sampling scheme of Baker (1987), preferred for its efficiency and lack of bias. A genetic operator is selected and applied to the sampled individuals according to a coin toss. "Mates" for the crossover operator are also selected from a fitness-weighted sample of the population. A final step was added to insure that the best individual from one generation is always retained in the next generation.

Good values for the genetic algorithm parameters are important to the efficient operation of the system. These parameters include the population size and the rates at which to apply the various genetic operators. The literature contains rules of thumb for setting the values of these parameters. For example, most previous work has employed a population size of 30 to 100 individuals; this seems empirically to be a good compromise between computational load, learning rate, and genetic drift.

Our modified crossover operator effectively exchanges homologous segments from the blueprints of two networks from the current generation to create a blueprint for a network in the next generation. In most applications of the genetic algorithm, homologous segments are identifiable by absolute positions in the bit string. For example, the Nth bit will always be used to specify the same gene in any individual. Because our representation allows variable-length strings, a modified two-point crossover operator (Booker 1987) was employed that determined homologous loci on two individuals by referring to the string's markers, discussed above.

The prime motivation for the variable-length representation (and hence the modified crossover) is to allow a much broader space of network architectures to be searched. For example, we do not want to fix a priori the number of areas or projections from an area. Simply encoding the "number of areas" as a field of a fixed-length bit string would not be sufficient, since each area also requires specification of other parameters such as the number of units, their configuration and so forth. We accommodate the possibility that the ideal network might have rather different specifications for each area and projection. Our blueprint representation allows specifications for areas and projections to be concatenated to produce arbitrarily complex network architectures.

The modified crossover is interesting from another standpoint. Self-similarity is sometimes an asset; specifications for areas or projections that have proved useful on one end of a network blueprint may prove useful on the other. The crossover operator, as defined above, can copy such segments en masse from one position in a blueprint to another, where, because of the nature of our representation, they have a fair chance of being exploited. Obvious parallels to the inversion operator can be drawn here. Applications of the genetic algorithm described in the literature have demonstrated an effective contribution from mutation at rates on the order of 10^{-2} or less. We have taken this as our setting.

Even though the bit string representation space was designed with closure under the genetic operators in mind, it is still possible for the genetic algorithm to generate individuals that are prima facie unacceptable. A common example would be a network plan that had no pathway of projections from input to output. Network blueprints may also arise that have areas with no projections, or projections that lead nowhere. An area that is not the target of any projection is similarly useless. Subtler problems arise from the limitations of our simulation capability. In our initial work we have abjured recurrence; network plans with feedback cannot be tolerated under the simple backpropagation learning rule. Two strategies have been

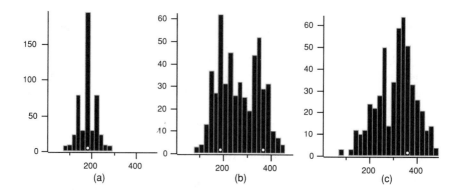

Figure 15.5: Frequency histograms showing the distribution of length (in bits) of the purified (see below) offspring produced with the modified crossover. Each histogram represents 500 crossover operations. The small white dots indicate the bins that would contain the parents. (a) Parent lengths 168. (b) Parent lengths 168 and 348. (c) Parent lengths 348.

employed for minimizing the burden of these misfits. First, the reproductive plan culls individuals with fatal abnormalities. Networks with no path between input and output areas compose the bulk of this group. Second, blueprints with minor abnormalities are "purified" in their network implementation, that is, their defects are not expressed. Our simulation software provides a switch that determines the fate of the unexpressed portions of an individual's chromosome. According to its setting, these "introns" may be retained in the blueprint or excised. We were initially concerned that the modified crossover, along with the purification process, would show uncontrollable biases toward either very large or very small individuals. Initial results suggest this is not the case. Figure 15.5 shows a typical distribution for the lengths of purified products of the crossover of same-length and differing-length parents.

A measure of fitness is necessary for the genetic algorithm to produce better and better networks. It is helpful to envision the algorithm as exploring the surface over the blueprint representation space defined by this function in an attempt to locate the highest peaks. The general plan for evaluation begins by purifying an individual's blueprint and instantiating the various units and weights indicated. As noted above, the blueprint does not represent weight values; the initial weights are set to small pseudorandom values. The individual network is then trained to criterion on the training set. Training alters the weights and thresholds of the network. The trained network may be then subjected to one or more tests in which, for example, stimuli are presented and responses noted, but the weights are not altered.

The attributes of networks are many and varied, and our relative interest in them may change from one application to the next. Consequently, we have adopted a melange of different performance and cost factors. The fitness function is a weighted sum of these performance metrics, p_i, optionally transformed by a function Ψ_i. The evaluation function, $F(i)$, for individual i is expressed in Equation 15.2.

$$F(i) = \sum_{j=1}^{n} a_j \Psi_j(p_j(i)) \tag{15.2}$$

The user of NeuroGENESYS can adjust the coefficients a_j to express the desired character of the network. Metrics that have been incorporated thus far include performance factors such as observed learning speed and the accuracy of the network on noisy inputs, and cost factors such as the size of the network in units and weights, and the maximal and average density of connections formed. The transformation, typically a sigmoid, serves to normalize or expand the dynamic range of the raw metric.

Because the relative weight on each metric can be modified, the network structure can be tuned for different optimization criteria. For example, if one of our goals is to synthesize networks that are computationally efficient, the size metrics might be given high weight. On the other hand, if accuracy and noise tolerance are more crucial, then the performance on noisy input patterns will be given a higher weight.

The metric chosen for learning rate requires some explanation. Because of limited computational resources, we cannot hope to train all networks until they achieve perfect accuracy on a given problem, or for that matter until any predetermined criterion is satisfied. A network may sometimes require a hundred epochs, while in others one hundred thousand may be insufficient. Training may be halted for a variety of reasons. We nonetheless wish to compare all individuals on the same learning rate scale, even though their training may have lasted different numbers of epochs and resulted in different final levels of accuracy. Our estimator is derived from integrating the rms error curve over the learning phase for the individual. The initial error is defined as the untrained accuracy of the network on the training set. The ratio of the area above the learning curve but below the initial error line to the area of the rectangle bounded by the initial error provides a unit-less measure of learning speed.

In order to make conclusions about the performance of the genetic algorithm (as opposed to the networks themselves) in discovering useful architectures, we require some standard to compare it against. Our approach is to run a control study in which network blueprints are generated at random, evaluated, and the best retained. This is effected simply by disabling the genetic operators of crossover and mutation.

IMPLEMENTATION

NeuroGENESYS is a Lisp program written at Honeywell's Sensor and System Development Center to provide a test bed for automated neural network design. The multi-window interface has a variety of control panels and displays for conducting experiments (Harp et al 1989b). It includes the ability to graphically inspect the areas and projections of individual neural networks and to modify them interactively.

The genetic algorithm offers a natural opportunity for speedup through parallelism. The NeuroGENESYS system runs on a network of Symbolics Lisp Machines. A single master machine manages the population, implementing the reproductive plan and controlling the other machines, whose role is to evaluate individual networks. The master also provides the interface for the user to set up and control experiments. The original Lisp network evaluation software has been rewritten in C and can also run on Apollo, Sun, and MicroVAX computers. These machines communicate with the master using the TCP/IP protocol.

The backpropagation network learning algorithm is notoriously slow on some problems. Both because backpropagation is computationally expensive and because a NeuroGENESYS experiment can involve training networks of widely different sizes on computers of varying performance, we have had to incorporate several limits on the computational resources devoted to evaluating any single individual. Training may be terminated for any of four reasons: (1) the problem has been learned to error criterion; (2) the maximum number of training epochs has been reached; (3) the maximum number of connection weight updates has been reached; and (4) the maximum training time has been reached. These limits are set on an experiment-by-experiment basis. In NeuroGENESYS, we have also implemented a recently discovered variation of backpropagation that often works significantly better for networks with multiple layers (Samad 1988).

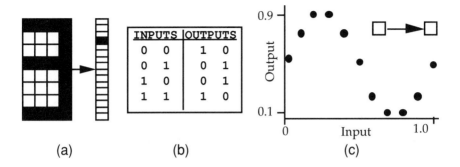

Figure 15.6: Typical stimuli from problems for which NeuroGENESYS has been used to optimize artificial neural networks. (a) One of the ten digits from the digit recognition problem. (b) A truth table for the two-bit exclusive OR problem. (c) The sine function approximation problem.

RESULTS AND ANALYSIS

We have tested the NeuroGENESYS software on a variety of problems that are typical of the sort which artificial neural networks trained by backprop-agation are used to solve. These include a simple digit recognition task, the "exclusive OR" problem, and an approximation of the sine function. Some typical stimuli from each of these problems are shown in Figure 15.6. Figure 15.6a shows the desired state of the output area for a typical input area state in the digit recognition problem. The input area is an 8 by 4 ar-ray of units, states of which can visually represent a digit; the output area is a vector of 16 units of which exactly one (indexing the input pattern) is to adopt the saturated state. (The length of the output area is 16 since the representation encodes dimensions as powers of 2.) For the exclusive OR problem, Figure 15.6b, the output units are to assume the exclusive OR function (and its negative) of the two input units; the outputs could be conceptually labeled as "even" and "odd" parity with respect to the inputs. Figure 15.6c graphs all 11 training examples, each indicating the desired real-valued output for a given real-valued input.

Most of our experiments were begun with a population of 30 individ-uals with architectures generated at random. Weights for the different performance measures were selected, and the system was set to run for a fixed number of generations. Our first experiment was with digit recog-nition, and NeuroGENESYS produced a solution that surprised us: The optimized networks had no hidden units but still managed to learn the

Indv 3152 on 2-Bit XOR

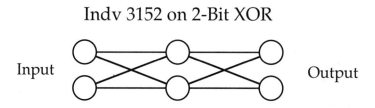

Figure 15.7: A typical architecture discovered for solving the exclusive OR problem with few connections.

problem perfectly. It had not been obvious to us that this digit recognition problem is linearly separable. Even in the simple case of no-hidden-layer networks, the value of application-specific design can be appreciated. When NeuroGENESYS was asked to optimize for average fanout as well as accuracy, the best network learned perfectly (though comparatively slowly) and had an average fanout of three connections per unit. With learning speed as the sole optimization criterion, the best network produced learned substantially faster (48 epochs), but it had an average fanout almost an order of magnitude higher.

Exclusive OR, of course, is the canonical non-linearly-separable problem. In this case, NeuroGENESYS produced many fast-learning networks that had a "bypass" projection from the input area directly to the output area in addition to projections to and from a hidden area. It is an as yet unverified conjecture that these bypass connections accelerate learning. The current blueprint representation scheme is not designed for working with very small networks; its stochastic routines for establishing the connections between units in projections work best with larger nets. Nonetheless, it is possible to find fairly small networks by making careful tradeoffs between accuracy, learning speed, and size. Figure 15.7 shows a typical network discovered by NeuroGENESYS for solving our version of the exclusive OR problem. In this experiment, accuracy was weighted at 0.75, while the number of connections measure had weight 0.25. For this problem, we believe this to be a nearly minimal configuration. (Note that our connection count does not include threshold weights; with threshold weights included, a one-hidden unit network with bypass connections would have one less connection.)

Figure 15.8 shows results obtained on the sine function approximation problem. Unlike the other problems, this one takes advantage of the analog nature of this class of artificial neural networks; only a single input unit and output unit were required. This problem required the network to operate in

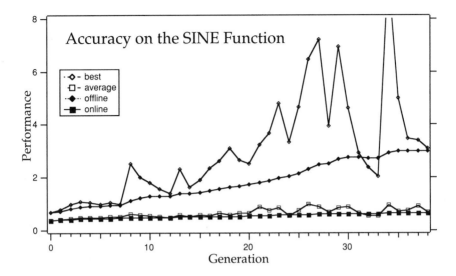

Figure 15.8: Performance on the sine function approximation problem.

the unsaturated region of its output unit activation function to approximate the sine. The fitness function for this experiment depended only on the accuracy of the trained network. In this case, improvements in online and average performance are overshadowed by dramatic improvements in the offline and best statistics.

In one of our experiments on the sine function problem, NeuroGENESYS was asked to design networks for moderate accuracy. The error cutoff during training was relatively high. Typically, the networks produced had one hidden layer of two units, which is the minimum possible configuration for a crude approximation. When the experiment was repeated with a low error cutoff, intricate multilayer structures were produced that were capable of modeling the training data very accurately. Figure 15.9 shows the architecture of a high-performing individual discovered for this problem.

NeuroGENESYS has consistently found better structures than those discovered by the same effort at random search (simulated by disabling the genetic operators). Appropriate network designs have been produced in a relatively small number of generations (less than 50).

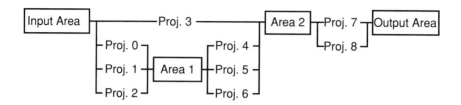

Figure 15.9: An individual that exhibited high accuracy on the sine function approximation problem.

FUTURE DIRECTIONS

The basic approach we have employed to explore artificial neural network architecture is extremely flexible. So far, the bulk of the experimental work done with this approach has employed the backpropagation learning rule and has been restricted to nets without recurrent projections. These restrictions have been useful in that this space of networks is comparatively well known and the computations required to train and test them are fairly simple (though time consuming). Through suitable alterations in the representation, the same genetic scheme can be applied to a wide range of artificial neural networks. We are currently investigating alternative neural network models employing other learning rules and architectures that allow recurrence. The synergy of evolution and learning offers a rich field for further exploration (Hinton and Nowlan 1986). It is no coincidence that this synergy itself has been successfully exploited by nature.

REFERENCES

Baker, J.E. (1987). Reducing bias and inefficiency in the selection algorithm. In J.Grefenstette (ed), *Proceedings of the Second International Conference on Genetic Algorithms*. Los Altos, Calif.: Morgan Kaufmann. pp. 14-21.

Bethke, A.D. (1980). *Genetic Algorithms As Function Optimizers*. Ph.D. thesis, University of Michigan.

Booker, L. (1987). Improving search in genetic algorithms. In L. Davis (ed.) *Genetic Algorithms and Simulated Annealing*. London: Pitman Publishers.

Caruana, R.A. and J.D. Schaffer (1988). Representation and hidden bias: Gray vs. binary coding for genetic algorithms. *Proceedings of the Fifth International Conference on Machine Learning*, pp. 153-161. Ann Arbor, Mich.

Edelman, G. (1988). *Neural Darwinism*. New York: Basic Books.

Harp, S., T. Samad, and A. Guha (1989a). Genetic synthesis of neural networks. Tech Report number I4852-CC-1989-2, Honeywell SSDC, 1000 Boone Avenue North, Golden Valley, MN 55427.

Harp, S., T. Samad, and A. Guha (1989b). Towards the genetic synthesis of neural networks. In J.D. Schaffer (ed.), *Proceedings of the Third International Conference on Genetic Algorithms*. Morgan Kaufmann.

Harp, S., T. Samad, and A. Guha (1990). Designing application-specific neural networks using the genetic algorithm. *Advances in Neural Information Processing Systems*, 2.

Hinton, G. and S. Nowlan (1986). How learning can guide evolution. Tech Report number CMU-CS-86-128, Computer Science Department, Carnegie-Mellon University, Pittsburgh, PA 15213.

Miller, G., P. Todd and S. Hegde (1989). Designing neural networks using genetic algorithms. In J.D. Schaffer (ed.), *Proceedings of the Third International Conference on Genetic Algorithms*. Morgan Kaufmann.

Montana, D.J. and L. Davis (1989). Training feedforward neural networks using genetic algorithms. *Proceedings of the Eleventh International Joint Conference on Artificial Intelligence*, pp. 762–767.

Purves, D. (1988). *Body and Brain*. Cambridge, Mass.: Harvard University Press.

Rumelhart, D. and J. McClelland, editors (1986). *Parallel Distributed Processing*. Cambridge, Mass.: MIT Press.

Rumelhart, D.E., G.E. Hinton, and R.J. Williams (1985). Learning internal representations by error propagation. ICS Tech Report 8506, Institute for Cognitive Science, University of California, San Diego.

Samad, T. (1988). Back-propagation is significantly faster if the expected value of the source unit is used for update. *Neural Networks* 1, Supp. 1: Abstracts of the First Annual INNS Meeting.

Schaffer, J.D., R. Caruana, and L. Eshelman (1989 unpublished). Using genetic search to exploit the emergent behavior of neural networks. Phillips Laboratories. 345 Scarborough Rd, Briar Cliff Manor, NY 10510.

Werbos, P.J. (1974). *Beyond Regression: New Tools for Prediction and Analysis in the Behavioral Sciences*. Ph. D. thesis, Department of Applied Mathematics, Harvard University, Cambridge, Mass.

Whitley, D. and T. Hanson (1989). Optimizing neural networks using faster more accurate genetic search. In J.D. Schaffer (ed.), *Proceedings of the Third International Conference on Genetic Algorithms*. Morgan Kaufmann.

Whitley, D. (1988). Applying genetic algorithms to neural net learning. Tech Report number CS-88-128, Department of Computer Science, Colorado State University.

16

Air-Injected Hydrocyclone Optimization via Genetic Algorithm

Charles L. Karr

INTRODUCTION

The air-injected hydrocyclone (AIHC) is a mineral-separating device recently developed by the U.S. Bureau of Mines (Stanley and Jordan 1989). This device combines the attributes of two individual separating devices: the selectivity of flotation and the simplicity of conventional hydrocyclones. In flotation, minerals are separated when selected particles are rendered hydrophobic (water-repellent) with chemical agents (called collectors) and attached to air bubbled through a slurry containing the minerals. These bubble-particle groups then rise to the top of the flotation cell where they are removed as a froth. Hydrophilic (water-attracted) particles are unaffected by the air and settle out separately (Sastry 1978). In conventional hydrocyclones, minerals are separated owing to large centrifugal forces created by a swirling flow characteristic of all hydrocyclones. These centrifugal forces cause large, high-density mineral particles to migrate toward the outer wall of the hydrocyclone, while small, low-density mineral particles move radially inward. The two groups of particles are removed from the hydrocyclone independently of each other. The basic design of an AIHC is shown in Figure 16.1. In the AIHC, the flotation process is enhanced by forcing it to occur in a centrifugal field where the centrifugal forces supplement the separating power of flotation.

The AIHC is an enhancement of the air-sparged hydrocyclone developed by Miller (1981). Although the optimization of AIHC design and operation has not been addressed in the literature, attempts to optimize air-sparged hydrocyclones have been ongoing since their introduction. Many of the published works concerning air-sparged hydrocyclone optimization have been produced by Miller, his students, and his co-workers (Kinneberg and Miller 1983; Miller, Misra and Gopalakrishnan 1986; Miller, Upadrashta, Kinneberg and Gopalakrishnan 1985; and Miller and Van Camp 1982). All these optimization studies have been experimental studies. However, the recent production of a mathematical model of an AIHC (Karr 1989) now makes it possible to use optimization methods for improving AIHC performance.

The function defined by the model for the predicted recovery of an AIHC is smooth and unimodal; thus, recovery can be maximized with a number of search techniques. In this chapter, the design and operation of an AIHC, as predicted by a mathematical model, is optimized using two search techniques: (1) a method described by Nelder and Mead (1965) and (2) a genetic algorithm (GA). The Nelder–Mead method is a robust search technique requiring no derivative information. Because of the well-behaved nature of the recovery predicted by the AIHC model, this method is well suited for this optimization problem. Like the Nelder–Mead method, the GA is a robust search technique (Goldberg 1989a). In the AIHC optimization problem, a three-operator GA with simplex reproduction locates a near-optimal AIHC after having sampled a number of points in the search space comparable to the Nelder–Mead method.

MATHEMATICAL MODEL OF AIHC

An AIHC typically consists of a conically shaped vessel with an opening at the apex to allow removal of the coarse or heavier particles (see Figure 16.1). The conical section is joined to a cylindrical section, the top of which is closed, with the exception of an overflow pipe known as a vortex finder. The vortex finder prevents the tangentially fed mineral from going directly into the overflow while allowing the fine particles a means of exiting the hydrocyclone. It is in this cylindrical section that the actual separation

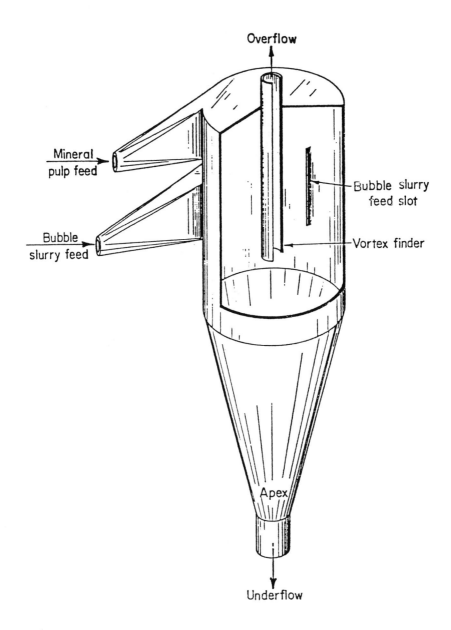

Figure 16.1: A schematic of an AIHC shows the external air injection. (Drawing courtesy of U.S. Bureau of Mines.)

occurs owing to the existence of a complex velocity distribution that carries the coarse particles to the apex and the fine particles out the top (Willis 1979). This cylindrical section is also characterized by the flotation process, made possible by the injected air and chemical treatment of the slurry feed, occurring in an AIHC.

Unlike the geometry of an AIHC, the mechanics of an AIHC are rather complex. A swirling flow is created when a mineral slurry is fed tangentially into the cylindrical section. This swirling flow creates centrifugal forces that cause the particles to migrate radially toward the outer wall of the hydrocyclone. The flotation process provides an additional means for separating particles. Although simple in principle, hydrocyclones provide effective separating capabilities that are further enhanced with air injection.

A mathematical model of an AIHC has been developed in order to demonstrate the effectiveness of GA-based design in the minerals industry (Karr 1989). The model is designed to predict the rate of flotation in the AIHC via three probabilities (Jameson 1977): (1) the probability of a bubble-particle collision (P_c), (2) the probability of the particle adhering to the bubble after collision (P_a), and (3) the probability of a bubble retaining an attached particle (P_r). P_c is estimated by solving the equations of motion for a single particle–single bubble system interacting in a strong force field (the centrifugal force field present in an AIHC). P_a is estimated by employing a model of a conventional hydrocyclone to find the d_{50} or split size, and thus the number of bubble-particle groups expected to leave with the overflow. At present, P_r is considered a function of the chemical reagents used in the flotation and is assumed to be unity.

This mathematical model provides an adequate indication of an AIHC's performance. Model calculations have been performed and compared to actual separation experiments at the Bureau of Mines, Tuscaloosa Research Center. A sample containing quartz and a sample containing phosphate were separated in independent tests. Results of model performance are shown in Figure 16.2 where the experimental recoveries of quartz and phosphate (the floated minerals) are plotted against the model-predicted recoveries. If the model agreed with the experimental results exactly, the points would fall on a 45° line.

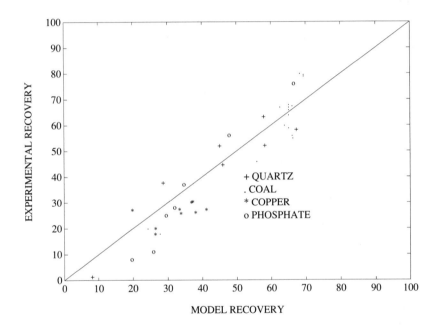

Figure 16.2: The mathematical model adequately predicts the performance of the AIHC.

OPTIMIZATION PROBLEM DESCRIPTION

Problem description is an important part of the optimization process. Initially, the decision variables considered in the optimization are defined. The variables considered are selected based on conventional hydrocyclone data, empirical studies of air-sparged hydrocyclones, and characteristics of the AIHC model. Next, constraints are placed on the decision variables. The constraints are selected based on physical and practical limits on hydrocyclone design and operation. Finally, an objective function is defined describing the goal of the optimization. The objective function is designed to produce an AIHC that provides maximum recovery while maintaining a minimum grade.

The first consideration in the optimization is to decide which parameters will be varied to alter AIHC performance. In this research, both design and operating parameters are considered because, to obtain optimal performance, a quality design must first be produced, and then the quality design must be utilized in the most efficient manner. If either of these classes of parameters was neglected, optimal performance could not be achieved.

There naturally are some constraints on each of the decision variables considered in the optimization study. These are generally practical constraints because of limitations in AIHC production methods and because of restrictions on support equipment (pumps, bubble generators, etc.) The limits on the parameters chosen in this study were selected so that, upon completion of the optimization, a working model of the optimized AIHC could be built and put into operation at the U.S. Bureau of Mines, Tuscaloosa Research Center, as part of a pilot plant. Therefore, C. E. Jordan (personal communication, November 20, 1988), project supervisor at the U.S. Bureau of Mines, suggested the limits. The parameters considered in this study and their respective limiting values appear in Table 1. One parameter that is considered but does not appear in the table is NBUB. A choice between two bubble generators, a fine bubble generator, and a coarse bubble generator, was available. NBUB, which stands for the number of the bubble generator, takes a value of 0 for the fine bubble generator and a value of 1 for the coarse bubble generator.

Because of the nature of the AIHC model, optimization of the design and operation of an AIHC must be performed for a specific mineral sample. In this study, an AIHC was optimized for the separation of a particular copper sample (Karr and Goldberg 1990). With the decision variables selected and a particular mineral sample specified, the objective function must be chosen. Selecting an objective function is at the heart of what constitutes a better AIHC.

Symbol	Parameters	Minimum Value	Maximum Value
D_c	AIHC diameter	1.2 cm	30.0 cm
D_i	Inlet diameter	0.06 cm	7.50cm
D_u	Underflow diameter	0.024 cm	4.500 cm
D_o	Vortex finder diameter	0.12 cm	9.00 cm
L	Vortex finder length	6.0 cm	40.0 cm
HD	Height-to-diameter ratio	5.0	20.0
GOS	Grams of solid	20.0 gm	200.0 gm
VOS	Volume of slurry	200.0 ml	500.0 ml
FR	Slurry feedrate	60.0 ml/s	2323.0 ml/s
PERWAT	Air-to-water ratio	50.0 %	90.0 %

Table 16.1: Parameters and Their Constraints

In this study, improved recovery is taken as the primary objective, where recovery is the percentage of mineral removed successfully with the overflow. However, recovery without satisfactory grade is unacceptable. Grade is the percentage of the desired substance relative to the total mineral removed in the overflow. Therefore, the objective of this study may be stated succinctly:

$$\left\{ \begin{array}{ll} \text{maximize} & R_r \\ \text{subject to} & G_r > G_{r_{min}} \end{array} \right.$$

Since unconstrained search techniques are used in this study, the following objective function (OBJ) is defined:

$$\text{OBJ} = \left\{ \begin{array}{ll} 0 & \text{if } G_r < G_{r_{min}} \\ R_r & \text{otherwise} \end{array} \right. \tag{16.1}$$

where OBJ is the objective function, G_r is the mineral grade achieved via the separation, $G_{r_{min}}$ is the minimum acceptable grade (28 percent), and R_r is the recovery achieved in the AIHC separation.

NELDER–MEAD OPTIMIZATION

The Nelder–Mead method is sometimes called the downhill (or uphill) simplex method because it uses a simplex of points to direct its movement toward an optimum. A simplex is a geometrical figure consisting, in M dimensions, of $M + 1$ points and all their connecting line segments and faces. In two dimensions, a simplex is a triangle, and in three dimensions, a simplex is a tetrahedron. The downhill simplex method utilizes an arbitrary simplex defined by the user and a number of reflection, contraction, and expansion rules to guide the procedure to the neighborhood of at least a local optimum. The method crawls downhill (for minimization, uphill for maximization) by renegotiating the position of the simplex in the search space until an optimum is found. A description of the method and a computer program implementing the method are found in Press, Flannery, Teukolsky, and Vetterling (1986).

In the previous section, the task of improving the operation of an AIHC was stated as a constrained optimization problem. The Nelder–Mead optimization method does not address constraints; therefore, some special consideration is required. Limits have been placed on each of the ten parameters, and the final solution must be one in which each parameter lies within its respective limits.

The search space considered by the Nelder–Mead method can be constrained in several ways. To ensure that the integrity of the method is maintained, a simple transformation of the variables is performed. Transformation of variables is an effective and common way of tackling constraints on decision variables (Box, Davies and Swann 1969). As an example of the transformation used in this research, consider the diameter of the AIHC, D_c. As the problem is now defined, D_c must lie between a minimum value, $D_{c_{min}}$ of 1.2 cm, and a maximum value, $D_{c_{max}}$ of 30.0 cm. When D_c is transformed according to

$$D_c = D_{c_{min}} + \left(D_{c_{max}} - D_{c_{min}}\right) \sin^2 D_c^* \qquad (16.2)$$

(where D_c^*, not D_c, is actually optimized), D_c will lie between the respective limits. This simple transformation insures that the constraint is met and maintains the integrity of the optimization method, yet it increases the required computations only slightly.

The Nelder–Mead method was used to find values of the ten parameters that maximized the recovery of an AIHC subject to a constraint on the grade. Table 2 shows the optimum values of these parameters, as selected by the Nelder–Mead method. Figure 16.3 illustrates the performance of the method. The recovery approaches a value of 82 percent after 290 function evaluations. Associated with this maximum recovery was the minimum grade of 28 percent. The 290 function evaluations serve as a standard against which the GA will be judged.

SIMPLEX REPRODUCTION

In this study, a three-operator GA consisting of reproduction, crossover, and mutation is used to optimize the design and operation of an AIHC. Before results of this optimization are presented and compared to those obtained in the Nelder–Mead optimization, a new form of reproduction that produces improved performance in the search of well-behaved functions is introduced. This reproduction scheme is known as simplex reproduction. Simplex reproduction characterizes an attempt to improve the convergence rate of a basic three-operator GA. This reproduction scheme utilizes small populations to reduce the number of function evaluations required to solve well-behaved optimization problems. A portion of the points in the small populations are generated at random in a fashion that resembles the projection of a simplex to discover new solutions in the Nelder–Mead method. In this section, we describe the simplex reproduction operator and provide a step-by-step procedure for its implementation.

The choice of parameters used in GA applications (population size, reproduction probability, mutation probability, etc.) is generally based on suggestions made in studies by De Jong (1975) and Grefenstette (1986). Both studies indicate that "bigger is better" when selecting a population size. When the population size is too small, premature convergence is generally obtained—the GA rapidly converges to a suboptimal solution. However, when the population size is too large, an excessive number of function evaluations is required, although near-optimal solutions are found.

In a recent work, Goldberg (1989b) describes an approach for the successful use of small populations in GA applications. This approach, implemented in the simplex reproduction operator, helps the GA overcome the shortcomings apparent in small-population GAs (i.e., limited information processing and premature convergence). The emphasis on limiting the population size arises from the desire to reduce the number of function evaluations required to achieve quality solutions. Based on this small population approach, Krishnakumar (1990) developed a microgenetic algorithm that used a population size of 5.

Simplex reproduction is an implementation of this fundamental approach. One strength of this small-population GA is that it carries its best solution in a given generation unchanged into the next generation, ensuring that the current best is at least as good as any solution found to that point (elitist reproduction). The mechanics of the simplex operator used to implement the small-population approach outlined above are summarized below.

1. Randomly generate N strings for the initial population, where N is a small odd number.

2. Evaluate the fitness of each string, select the best solution based on fitness, place it in position N, and send it unchanged into the next generation.

3. Force strings that are adjacent to each other in the population to compete directly with each other for the right to survive. Similarly, the more fit string between strings 1 and 2 is placed in the mating pool. The more fit string between strings 3 and 4 is placed in the mating pool. The local competitions are held until strings $N - 2$ and $N - 1$ compete. The $\frac{(N-1)}{2}$ strings in the mating pool are crossed with a probability of 1.0 and placed in the next generation. The local competitions enhance the Darwinian "survival-of-the-fittest" aspect of the GA. They help ensure that the best solutions thrive in the population.

4. At this point, there are $\frac{(N-1)}{2} + 1$ strings in the next generation. The remaining $\frac{(N-1)}{2}$ strings are generated in a random fashion. Actually, the strings are formed by mutating the best string with a probability of p'_{mutate}.

5. Continue the cycle initiated in step 2 until convergence is achieved. The population is considered converged when none of the strings in the population differs from the population's current best solution by more than two bits in a bit-by-bit comparison.

6. Take the best solution in the converged generation and place it in a second "initial generation." Generate the other $N - 1$ strings in this second initial generation at random, and begin the cycle again until a satisfactory solution is obtained.

Basically, this simplex reproduction operator amounts to starting with some initial population, getting the most possible out of the initial population, and then using that initial population's greatest potential solution as the starting point for another search. Simplex reproduction is, in a sense, an attempt to provide the GA with hill-climbing capabilities. However, it should not be viewed as an attempt to convert the GA into a hill climber.

As noted earlier, a major drawback of small populations is their tendency to converge prematurely. Simplex reproduction recognizes and allows for this tendency in that when premature convergence is achieved, the GA simply uses its most valuable information to date (its best solution) as a starting point for a new attempt to find a near-optimal solution.

SIMPLEX GENETIC ALGORITHM OPTIMIZATION

In this section, we use a three-operator GA employing simplex reproduction (simplex GA) to optimize the design and operation of an AIHC. The results obtained are compared to those of the Nelder–Mead method and a GA with roulette wheel selection.

A computer program implementing the simplex GA was written in Pascal and used to optimize the design and operation of an AIHC. The following characteristics were used in the study:

popsize	21
$p_{mutation}$	0.05
$p_{crossover}$	0.85
p'_{mutate}	0.3
string length	51
coding	concatenated, mapped, unsigned binary
selection method	simplex
scaling	linear

The performance of the simplex GA is again illustrated in Table 2 and Figure 16.3. As seen in Table 2, the values of the parameters suggested by the simplex GA are similar to those suggested by the Nelder–Mead method and by a GA using roulette wheel reproduction. Figure 16.3 shows that the recovery again approaches a value of 82 percent with 28 percent grade. The inclusion of the simplex reproduction scheme has improved GA performance to a point where it is comparable to that of the Nelder–Mead method.

Symbol	Nelder-Mead	GARW	Simplex GA
D_c	26.213 cm	26.284 cm	26.284 cm
D_i	7.052 cm	7.004 cm	7.004 cm
D_u	3.341 cm	3.606 cm	3.606 cm
D_o	1.526 cm	1.304 cm	1.304 cm
L	28.213 cm	27.933 cm	29.030 cm
HD	10.240	10.324	10.324
GOS	196.018 gm	197.143 gm	197.143 gm
VOS	400.252 ml	404.760 ml	404.760 ml
FR	2241.337 ml/s	2233.905 ml/s	2216.086 ml/s
PERWAT	86.251 %	87.333 %	87.333 %
NBUB	0	1	0

Table 16.2: Best Values as Selected by the Nelder–Mead method, a GA with Roulette Wheel Reproduction (GARW), and a Simplex GA

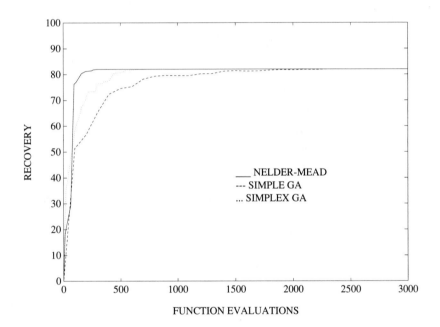

Figure 16.3: The simplex GA locates near-optimal solutions rapidly.

SUMMARY

The design and operation of an AIHC were optimized using a genetic algorithm. Results were compared to those obtained using a standard search technique, the Nelder–Mead method. Each method converged to approximately the same values for the ten decision variables considered in this study. Each converged to a maximum recovery of 82 percent with the minimum allowable grade. The U.S. Bureau of Mines is currently developing an AIHC based on the suggestions of this study.

The Nelder–Mead optimization method was used initially to find an improved operating point for the AIHC. After the optimization problem was completely defined, the Nelder–Mead method was applied and converged to an improved solution. The performance of this method was used as a standard against which the simplex GA was judged.

A simplex genetic algorithm was used as a second method for optimizing AIHC design and operation. First, the simplex GA was briefly described. Next, a simplex GA was applied to the AIHC design and operation optimization problem. The results were similar to those found by the Nelder–Mead method. The introduction of this small-population GA should enhance the performance of the already powerful GA in well-behaved optimization problems.

The function defined by the AIHC model is smooth and well defined; therefore, the Nelder–Mead method is "tailor-made" for this optimization problem. With the introduction of simplex reproduction, which is an attempt to provide the GA with more direct hill-climbing capabilities, a three-operator GA was able to nearly match the performance of the Nelder–Mead method. The simplex GA should improve GA performance on problems that are smooth and well behaved. It is not likely they will be able to out perform hill climbers on every problem of this type. However, they should improve GA performance on problems that are locally smooth and well behaved, as well as on problems that give hill climbers problems for any reason.

Certainly, the GA has not been used in this study to do anything new and different; that was not the intent. This study was performed to demonstrate the GA's ability to perform efficiently on a real-world optimization problem and to perform comparably with an effective search method. Those goals have been accomplished.

REFERENCES

Box, M. J., D. Davies, and W. H. Swann (1969). *Non-Linear Optimization Techniques.* Edinburgh: Oliver and Boyd.

De Jong, K. A. (1975). *Analysis of the Behavior of a Class of Genetic Adaptive Systems. Dissertation Abstracts International,* vol. 36, 5140B, University Microfilms no. 76-9381.

Goldberg, D. E. (1989a). *Genetic Algorithms in Search, Optimization, and Machine Learning.* Reading, Mass.: Addison-Wesley.

Goldberg, D. E. (1989b). Sizing populations for serial and parallel genetic algorithms. *Proceedings of the Third International Conference on Genetic Algorithms and Their Applications* 3, pp. 70–79.

Grefenstette, J. J. (1986). Optimization of control parameters for genetic algorithm. *IEEE Transactions on Systems, Man, and Cybernetics* 12(16), p. 122.

Jameson, G. J. (1977). Physical factors affecting recovery rates in flotation. *Minerals Science and Engineering* 3(9), p. 103.

Karr, C. L. (1989). A model of recovery in an air-sparged Hydrocyclone. *Proceedings of the International Association for Hydraulic Research XXIII Congress* S, pp. 1–8.

Karr, C. L. and D. E. Goldberg (1990). Genetic algorithm based design of an air-injected hydrocyclone. In R. Rajamani and J. A. Herbst (ed.), *Proceedings of Mineral and Metallurgical Processing Symposium, Control '90.* Salt Lake City, Utah: AIME, pp. 265-272.

Kinneberg, D. J. and J. D. Miller (1983). Copper sulfide flotation in an air-sparged hydrocyclone. Open File Report 149–83, U.S. Bureau of Mines, Tuscaloosa, Ala.

Krishnakumar, K. (In press). Microgenetic algorithms for stationary and non-stationary function evaluation. *Advances in Intelligent Robotics Systems,* Philadelphia: SPIE.

Miller, J. D. (1981) Air-sparged hydrocyclone method. *U.S. Patent 4,279,743.* July 21, 1981, Washington, D.C.

Miller, J. D., M. Misra, and S. Gopalakrishnan (1986). Fine gold flotation from Colorado river sand with the air-sparged hydrocyclone. *Minerals & Metallurgical Processing,* 3(3), pp. 145–148.

Miller, J. D., K. R. Upadrashta, D. J. Kinneberg, and S. Gopalakrishnan. (1985). Fluid-flow phenomena in the air-sparged hydrocyclone. *International Mineral Processing Congress,* Vol. 15, pp. 87–99.

Miller, J. D. and M. C. Van Camp (1982). Fine coal flotation in a centrifugal field with an air-sparged hydrocyclone. *Mining Engineering,* 3(34), pp. 1575-1580.

Nelder, J. A. and R. Mead (1965). Simplex method for function minimization. *Computer Journal,* 7, pp. 308–318.

Pess, W. H., B. P. Flannery, S. A. Teukolsky, and W. T. Vetterling (1986). *Numerical Recipes.* Cambridge: Cambridge University Press.

Sastry, K. V. S. (1978). Flotation of mineral fines. *Report of a Workshop, August 27–29,* Sterling Forest, N. Y.

Stanley, D. A. and C. E. Jordan (1989). Bubble-injected hydrocyclone flotation cell. *U.S. Patent pending, application no. 336,168.* U. S. Patent Office, Washington, D.C.

Willis, B. A. (1979). *Mineral Processing Technology.* Elmsford, N. Y.: Pergaman Press.

17

A Genetic Algorithm Approach to Multiple-Fault Diagnosis

Gunar E. Liepins and W. D. Potter

INTRODUCTION

In communications as well as medicine and other domains, multiple fault-diagnosis is the identification of a set of conditions (faults or diseases) that best corresponds to or explains some observed abnormal behavior (alarms or symptoms) (Peng and Reggia 1987a; Reiter 1987). Because of the importance of these types of problems and the difficulty humans experience in dealing with them quickly, automating multiple-fault diagnosis has been the focus of extensive research efforts (Davis 1984; Genesereth 1984; Josephson et al. 1987; de Kleer and Williams 1987; Peng and Reggia 1987a and 1987b; Reggia et al. 1983; Reiter 1987).

This chapter describes a genetic algorithm-based diagnostic system for diagnosing multiple microwave communication faults. The system has been developed to augment the Communication Alarm Processor Expert System (CAP) developed at Oak Ridge National Laboratory for the Bonneville Power Administration (Potter et al., in press, b; Purucker et al. 1989). The CAP system was designed to perform single-fault diagnosis using communication alarms as indicators of abnormal behavior of the microwave network. CAP is not readily extendable to multiple faults because of the knowledge engineering approach taken during its original development. (More detail on these limitations is presented later.) Previous approaches to multiple-fault diagnosis use standard connectionist techniques (Peng and Reggia 1989) or traditional artificial intelligence techniques (de Kleer and Williams 1987; Potter et al., in press, b). These other studies notwithstanding, multiple-fault diagnosis remains a difficult problem (in most cases an NP-Complete problem (Garey and Johnson 1979)).

The multiple-fault diagnosis problem is naturally a 0–1 integer programming problem. Special classes of 0–1 programming problems have efficient operations research and mathematical programming solution techniques. However, the general multiple-fault diagnosis problem does not apparently fall into any of these special classes. Because the fault diagnosis problem often has a nonstandard operations research formulation and because microwave communication diagnosis must be near real-time, the aforementioned approaches are unlikely to be successful. Genetic algorithms have been demonstrated to solve poorly characterized 0–1 programming problems efficiently (Liepins et al. 1990). Hence, they were selected for this application.

Multiple-fault diagnosis problems range from the very difficult to the still more difficult. Difficulty can be measured along two dimensions: static versus dynamic and deterministic versus probabilistic. The easiest are problems with a fixed discrete set of alarms and deterministic causality between faults and alarms. That is, each set of faults deterministically triggers an associated set of alarms that remain unchanged during the time required for diagnosis. More difficult are problems in which the relationship between faults and alarms is stochastic; alarms themselves may be faulty. Still more difficult are stochastic problems in which different alarms (associated with any given fault) arrive at different times, new faults can occur continuously, and repairs can be made continuously. These latter problems arise in microwave communications diagnosis.

Solution techniques for near real-time, dynamic, stochastic diagnostic problems need to be fast and need to be able to make use of partial alarm information. At any given step in the diagnosis process, the current tentative diagnosis may need to be refined or replaced owing to the triggering of additional alarms. This is similar to *nonmonotonic* reasoning in artificial intelligence where conclusions may be rejected in the face of additional information.

A set of common cause faults may be somewhat staggered in their triggering of alarms. Conversely, unrelated faults may occur at the same time and may trigger (near) simultaneous alarms. The degree to which such "alarm mixtures" increase the difficulty of diagnosis is not well understood. The proposed adaptations of the genetic algorithm to handle these cases are described later in this chapter.

SOLUTION APPROACHES

At least four generic approaches seem plausible for multifault diagnostic problems: (1) knowledge engineering, (2) learning from actual examples, (3) learning from model-generated examples, and (4) directly optimizing models. Each has advantages and disadvantages. The current extension to the CAP system is based on the fourth approach, directly optimizing a model.

The first approach to diagnosis, direct knowledge engineering, requires encoding diagnosticians' experiences into rules. Moderate-sized, relatively static systems that have been monitored extensively lend themselves to a knowledge engineering approach. The diagnosticians are likely to be able to articulate the reasoning and rules they use to perform diagnoses. To be practical, any resultant knowledge base would be limited to no more than a few thousand rules; knowledge engineering is typically a slow and expensive process, and larger knowledge bases would be prohibitively expensive. We should expect the usual difficulties with knowledge engineering, such as difficulties with updating and insuring completeness, consistency, and correctness.

When a large set of alarm-diagnosis pairs is known, but look-up tables are inadequate and diagnostic processes are not well understood, machine induction from examples might be appropriate. For example, look-up tables are inadequate if only a subset of the correct diagnoses is currently known or if the look-up tables are overly large. Knowledge engineering fails if each alarm incident seems to require a specialized line of reasoning, or if diagnostic procedures can be articulated only for a small sample of alarm combinations. Approaches to machine induction from examples include decision trees (Quinlan 1986), neural nets (Jacobs 1988), and classifier systems (Goldberg 1989). In cases where the underlying structure has considerable regularity, these methods are useful and can discover rules and representations equal (and sometimes superior) to those articulated by experts (Michalski 1987). On the other hand, if the underlying structure and "natural" representations are especially pernicious, these methods cannot be expected to produce highly useful results (Valiant 1984). In these instances, machine induction does not necessarily discover concepts, and any concepts that have been discovered may generalize poorly. (For a limited performance comparison of decision trees and neural nets, see Mooney et al. (1989)).

A further level of difficulty may occur if neither diagnostic processes nor a large number of alarm-diagnosis pairs are known. In this case a modeling approach may still work. Depending on the complexity and computational requirements of the model, one of two tacks can be taken. If the model's computational requirements exceed the maximal system response time, the model can be used off-line to generate alarm-diagnosis pairs and machine induction used with these pairs to generate rules. All the difficulties of learning from historically derived examples also apply to the case of model-generated examples. In addition, the issue of model faithfulness needs to be addressed. For each set of alarms, does the model generate the proper diagnosis? Model validation of this sort is far from trivial.

The other alternative is to develop a model that can be used *on-line*. This is exactly what was done for the CAP system (Potter et al., in press, b). The elementary model discussed below serves to introduce the more complicated model actually used. Each model is discussed from the perspective of its faithfulness (to the actual phenomenon) and its genetic algorithm solution.

SET-COVERING MODELS

A conceptually plausible model for diagnosis is the set-covering model. Set covering for diagnosis has been investigated by Garfinkel, Liepins, and Kunnathur (1986 and 1988). Set-covering problems are a class of easily stated but provably difficult problems that have long been studied by operations researchers (Garey 1979). Many algorithms have been suggested for their solution. Two well-known algorithms were developed by Bellmore and Ratliff (1971) and Balas and Ho (1980), respectively. Genetic algorithms appear to offer an attractive alternative means for solving set-covering problems (Liepins et al. 1990).

In the framework of the CAP system, consider a matrix of 0–1 column vectors each of which represents a fault. A "1" in the column represents an alarm that is triggered by the fault, whereas a "0" in the column signifies that the corresponding alarm fails to be triggered. (In this model each fault deterministically triggers a set of alarms. Probabilistic models are discussed below.) Each fault i has some prior probability p_i of occurring.

The cost $c_i = 1/p_i$ is associated with the ith column. Let K be a k x n matrix of 0's and 1's, whose columns are the fault-alarm vectors. Let c be an n-dimensional row vector of nonnegative costs (related to probabilities as above). Assume that m alarms have been activated. Let M be the m x n submatrix of K determined by the rows of K that correspond to the activated alarms. Let x be an n-dimensional column vector of indicator variables. That is, each component x_i of the vector x is either 0 or 1. The formal set-covering diagnostic problem becomes

$$\min_{x} \ \sum c_i x_i$$
$$\text{subject to } Mx \geq 1,$$
$$x_i = 0, 1$$

where inequality is defined to be inequality of vectors, that is, component-by-component inequality. Conceptually, this can be interpreted as "find the set of faults with maximum probability which would trigger the observed alarms." (This conceptual interpretation can be formally justified if independence of faults is assumed; see, for example, Peng and Reggia 1987a and Liepins and Pack, in press).

SET-COVERING SOLUTIONS

The set-covering problem is a constrained optimization problem. Normally, genetic algorithms are directly applicable only to unconstrained problems. Constrained optimization problems cause some potential difficulty for genetic algorithm implementations. Consider the following example with

$$cost = 3x_1 + 2x_2 + 3x_3 + 2x_4 + 2x_5 + x_6$$

and

$$M = \begin{pmatrix} 1 & 0 & 0 & 0 & 0 & 1 \\ 0 & 1 & 1 & 0 & 0 & 0 \\ 0 & 1 & 1 & 0 & 0 & 1 \\ 1 & 0 & 0 & 1 & 1 & 0 \\ 0 & 0 & 1 & 0 & 1 & 0 \\ 1 & 0 & 0 & 1 & 0 & 0 \end{pmatrix}$$

The string (0 0 1 1 1 1) indicates that the third column is used in a candidate solution, the first column is not, and so forth. If we have two feasible candidate solutions, recombination of the two need not be feasible. That is, both candidate solutions may be covers, but their recombination need not be. Consider, for example, the two strings a and b crossed between the second and third bit position to generate the string c by concatenating the first part of a with the last part of b. Both a and b represent covers, but c does not.

$$
\begin{array}{ccccccccc}
a & = & 0 & 0 & | & 1 & 1 & 1 & 1 \\
b & = & 1 & 1 & | & 0 & 0 & 1 & 0 \\
c & = & 0 & 0 & & 0 & 0 & 1 & 0
\end{array}
$$

At least three genetic algorithm formulations for constrained optimization problems have been frequently mentioned in the literature. First, invent specialized recombination operators that maintain feasibility. Second, throw away infeasible solutions and repeat recombination until a feasible solution is generated. Third, allow infeasibility, but adapt a penalty function so that the cost of a *solution* is the sum of the costs of the columns used plus a penalty if the string represents an infeasible solution. Only the first and third alternatives have been shown to be useful. Which alternative is best depends on the particular problem and domain involved. Problems such as the traveling salesman problem, for which most solutions in the "natural" representation are not feasible, seem to be best attacked by a change of representation or modified recombination operator. For other problems, appropriately modified recombination operators are not known, and penalty function approaches have to be taken. If a penalty function approach is selected, the choice of the penalty significantly affects genetic algorithm performance. The rule of thumb that has emerged is that good penalty functions are those that provide an easily calculated "tight estimate of the cost of completion." For set-covering problems, consider the two penalty functions P_1 and P_2 below: Let N be the set of indices $N = \{1, \ldots, n\}$ and S index the columns in the current solution. If S fails to be a cover, set

$$
P_1 = \sum_{N-K} c_i
$$

The definition of P_2 is somewhat more complicated. Let R be the rows that remain uncovered. For each $i \epsilon R$, let S_i be the set of columns that cover row i. Let $c_i* = \min\{c_j\}$ for $j \epsilon S_i$. Then

$$
P_2 = \sum_{R} c_i*
$$

The function P_1 fails to discriminate between different types of infeasible solutions—"near" infeasibility is treated no better than "far" infeasibility. The function P_2 discriminates between these two cases. Liepins (1990) demonstrates the superiority of P_2 over P_1 and suggests that with P_2 the genetic algorithm approach to the solution of set-covering problems seems to be competitive with specialized operations research approaches. Richardson (1989) provides additional evidence to support the emergent rule of thumb about the relative merit of alternative penalty functions. (In some sense, this rule of thumb is the genetic algorithm counterpart to the artificial intelligence search result for the $A*$ algorithm (Pearl 1984). Whereas the $A*$ algorithm requires an optimistic heuristic function to guarantee optimality, the genetic algorithm requires a "pessimistic cost of extension." A counterpart observation has been made by Glover (1988) for TABU search. Glover observes that it is useful to search from outside as well as from within the feasible region.)

EXTENDED MODELS

Three objections can be raised against the set-covering approach to multiple-fault diagnosis. First, it might be argued that the model is overly simplistic, that it does not reflect the full complexity of the disambiguation process. Second, the model may (and in the case of equal prior probabilities, sometimes does) determine multiple alternative optimal solutions. Third, even if the model were faithful and the optimal diagnosis were unique, it is possible that any set of alarms was actually triggered not by the most likely event consistent with the constraints, but by some other less likely event. In the case of many possible, consistent explanations differing only slightly in their probabilities, identification of the most likely cause can often result in the wrong diagnosis, even if the model is perfectly faithful (Liepins and Pack, in press; Weyrich et al. 1990). This problem will henceforth be called lack of model selectivity.

The first objection can be overcome by refining the model; such a refined model was actually used for the CAP system (Potter et al. in press, b). The second problem, the problem of multiple optima, rarely occurs in the case of nonequal prior probabilities, and in the case of equal probabilities, it can be overcome with use of an auxiliary discriminant function. In contrast, the third problem, that of many possible diagnoses differing only slightly in their probability, is inherent in the system; no nonomniscient diagnostic procedure can do better than to use all available information to suggest the most likely diagnosis and list the remaining diagnoses in the order of their probabilities (perhaps dynamically modified as additional tests are performed). Any improvement in performance would require a more discriminating set of alarms.

An extension to the set-covering diagnosis model that captures stochastic fault-alarm relationships replaces the 0–1 fault-alarm column vectors with real valued column vectors. Thus, the entry p_{ij} in the ith row of the jth column gives the conditional probability that the jth fault triggers the ith alarm given that the jth fault has occurred. Simultaneously, the constraints can be modified to reflect the condition that faults that normally trigger alarms outside the current observed set should not be part of the diagnosis. Let I index the observed alarms, and let K be the k x n fault-alarm probability matrix. As before, let M be the submatrix of K indexed by the rows of I. Let R be the submatrix of the remaining rows. Let $0 < t1 < t2 \leq 1$ be two thresholds. Then the diagnosis problem can be formulated as

$$\min \sum c_i x_i$$
$$\text{subject to } Mx \geq t2,$$
$$Rx \leq t1,$$
$$\text{and } x_i = 0, 1$$

This model is very similar to that of Peng and Reggia (1987a; 1987b), a variant of which was used for the CAP system (Potter et al., in press, b). For any set of observed alarms A, Peng and Reggia determine the diagnosis to be the set S of faults that maximizes the product $L(S, A) = L1 * L2 * L3$ where

- $L1$ is the probability that the alarms A are observed given the set S of faults

- $L2$ is the probability that alarms not in A are not observed given the set S of faults

- $L3$ is the prior probability that the faults S (and no others) are present.

The Peng and Reggia (1987a) model differs from the extended set-covering model in two important ways. (1) The matrix inequalities used to check compliance with the threshold values cannot be shown to be the probabilities $L1$ and $L2$ except under special conditions. On the other hand, they can easily be replaced by the terms $L1$ and $L2$, respectively. (2) Instead of incorporating the terms $L1$ and $L2$ into the objective function, the extended set-covering formulation requires that thresholds be satisfied by the counterparts to these terms.

IMPLEMENTATION

Similar to the argument that suggests a graded penalty function for constrained optimization, it is useful to modify the Peng and Reggia model to provide feedback to search algorithms whenever the strict model returns the value zero. The strict model does so whenever either $L1$ or $L2$ is zero, that is, whenever the candidate solution fails to be a cover or includes a fault that has a causal association with an alarm not present in the active alarm set. One modification that works is to set a nonzero lower bound for each of $L1$ and $L2$. Because the product involves three terms, this allows a gradation in the fitness values even for noncovers and extraneous covers. (*Without* this modification to the model, the genetic algorithm often failed to find optimal solutions in our tests. In fact, the genetic algorithm used with the strict model resulted in optimal solutions less than 10 percent of the time. The modified model resulted in 85 percent or better optimal solutions (Potter et al., in press, a; in press, b).

The current genetic algorithm implementation uses two-point recombination, population sizes varying between 50 and 150, a recombination probability of 0.6, and a mutation probability of 0.0333. Other technical implementation details that were explored are described below. Of these, the elitist policy and seeding were tried but had little impact on the overall reliability and so are not currently used. Mass mutation and post processing proved useful and are being employed.

Elitist policy: If no increasingly fit individual has been discovered between generations, the elitist policy simply carries forward the most fit individual from the previous generation into the next. (In other optimization problems, this policy has sometimes led to premature convergence.)

Seeding: Seeding inserts into the initial population a single individual who represents the presence of all those faults that have a causal relationship to any of the active alarms and the absence of all other faults. The other individuals are randomly generated. Thus, at least one individual in the initial population is guaranteed to be a cover. (In other studies, Booker (1987) has found that thoroughly randomizing the initial population improved genetic algorithm results. Liepins (1990) found little benefit to seeding the initial populations of traveling salesman problems with locally optimal tours.)

GA	10x10	10x12	10x15	10x20
(50 no post)	60.51	78.59	81.23	70.19
(100 no post)	75.86	90.71	91.10	83.28
(150 no post)	81.92	95.01	96.87	89.74
(50 post)	85.24	99.02	99.41	98.34
(100 post)	90.22	100.00	99.90	98.53
(150 post)	91.30	99.71	99.90	99.32

Table 17.1: GA performance given as a percentage of Optimal solutions found

Mass mutation: Mass mutation represents an attempt to adapt the genetic algorithm to dynamically changing problems. If the genetic algorithm is used to solve a sequence of unrelated problems, then clearly the solution to one has no bearing on the solution of another. A reasonable strategy would be to restart the genetic algorithm with a newly generated population (often randomly generated) independently of the previous solution. In the alarms environment where alarms are triggered or resolved with varying interarrival/service times, it is advantageous to use the evolving solution populations from one problem to "seed" the search for the next solution. (Parsing remains an admitted difficulty. When is the next alarm configuration closely related to the previous one?) The alternative that suggests itself is to choose as an initial population for the incoming problem some (current or earlier) population of the evolving current solution modified by (mass) mutations.

Post processing: The post-processing scheme takes the genetic algorithm-produced solution and systematically adds and deletes single faults in the diagnosis. This is effectively hill climbing starting from the genetic algorithm-produced solution. Whether or not it improves the CAP's solutions depends on the relationships between the faults and the alarms. In cases where each fault triggers few alarms and each alarm is triggered by no more than a few faults, post processing seems to do little to improve solutions. However, in randomly generated, high-redundancy relationships, post processing significantly improves the solutions, as can be seen in Table 17.1. These results are not altogether surprising insofar as post-genetic algorithm hill climbing has been invoked for other problems with varying success (Liepins et al. 1990).

Sensitivity analysis: Another post-processing procedure is more properly called sensitivity analysis. Similarly to post-genetic algorithm hill climbing, each fault is systematically deleted from the diagnosis and the percentage change in the diagnosis fitness is calculated. This sensitivity analysis helps to prioritize the faults and provides information about central common faults across dynamically changing sets of alarms.

DATA AND RESULTS

An obvious necessity for a model-based approach to fault diagnosis is to accurately know the model parameters. The modified Peng–Reggia model requires the prior probabilities of the faults and the causal association matrix that prescribes the probabilities that any specific alarm is triggered by a given fault, given that the fault is present. These probabilities are not well known for the current CAP system. To date, 19 possible faults and 41 alarm groups have been identified, but no failure statistics for the faults have been accumulated. The prior fault probabilities have been arbitrarily set to $p = 0.5$ until enough historical data on the mean time between failures can be collected and incorporated into the system. (This is simply the uninformed prior; all faults are deemed equally likely. Moreover, the choice of 0.5 is purely arbitrary.)

The system was tested at two levels. First, the genetic algorithm solutions were compared to model optima. Second, the genetic algorithm was tested with known alarm sets from the CAP system. Solutions were informally compared with human expert diagnoses. Personnel at the Bonneville Power Administration subjectively judged the results to be good. (The second set of tests evaluated genetic algorithm optimization, model faithfulness, and model selectivity jointly. Individual factors were not separated out.) For the optimization tests for cases with randomly generated relationships between faults and alarms (dense causal association matrices), genetic algorithm optimization performance with and without post processing is given in Table 17.1. The results reflect genetic algorithm performance averaged over one case of each of the 1023 possible alarm combinations. The statistic is the percentage of cases for which the genetic algorithm solution equaled the optimal. (The parenthetical entries give the population sizes and show whether or not post processing was employed. The first number of the column heading pairs [10] specifies the maximum number of alarms. The second number specifies the total number of components.) In the worst case (genetic algorithm population size set to 50), the genetic algorithm with post processing fully optimized about 85 percent of all the possible alarm sets and, in all but 2 percent of the cases, found either the optimal, penultimate, or third best solution.

To date, no direct comparisons of the genetic algorithm approach (to the solution of the Peng–Reggia model) have been made with other approaches. Previously published results for the connectionist techniques used by Peng and Reggia (1989) suggest that for the 10 x 10 case, genetic algorithm and connectionist techniques are roughly comparable in terms of solution quality. No comparisons are available for CPU time or for the larger problems.

The future of the genetic algorithm enhancement to the CAP system is uncertain at this time; the Bonneville Power Administration is well pleased with the results to date, but full implementation depends on additional funding and on compiling of the data required for the association matrix (which represents the actual behavior of the microwave communication system). The status of both remains unknown.

SUMMARY

Four approaches to multiple-fault diagnosis were introduced and discussed: knowledge engineering, learning from actual examples, learning from model-generated examples, and directly optimizing models. Each was presented to have its advantages and drawbacks. Three formal modeling approaches were presented: set covering, extended set covering, and the modified Peng and Reggia (1987a) model. The first two of these result in constrained optimization problems. Penalty function approaches to improve genetic algorithm solutions for constrained optimization were reviewed. Selected results of genetic algorithm optimization of the modified Peng and Reggia (1987a) model for the CAP system were cited. These results, together with those of Liepins (1990), suggest that genetic algorithms frequently provide optimal or near optimal solutions to a variety of (small) multiple-fault diagnosis formulations, including set covering and the modified Peng and Reggia (1987a) model. For the cases studied, post processing seems to be a computationally inexpensive means of improving solutions. To date, these conclusions can only be drawn for small problems. How genetic algorithms perform as problem size increases remains to be investigated.

REFERENCES

Balas, E., and A. Ho (1980). Set covering algorithms using cutting planes, heuristics, and subgradient optimization: a computation study. *Mathematical Pro-*

gramming 12, pp. 37–60.

Bellmore, M., and H. D. Ratliff (1971). Set covering and involuntary bases. *Management Science* 18, pp. 194–206.

Booker, L. (1987). Improving search in genetic algorithms. In Davis (ed.), *Genetic Algorithms and Simulated Annealing*. London: Pitman, Los Altos, Calif., pp. 61–73.

Davis, R. (1984). Diagnostic reasoning based on structure and behavior. *Artificial Intelligence* 24 (1–3), pp. 347–410.

Garey, M. R., and D. S. Johnson (1979). *Computers and Intractability: a Guide to the Theory of NP-Completeness*. San Francisco: Freeman Publishers.

Garfinkel, R. S., A. S. Kunnathur, and G. E. Liepins (1988). Error localization of erroneous data: continuous data, linear edits. *SIAM Journal of Scientific and Statistical Computing* 9(5), pp. 922–931.

Garfinkel, R. S., G. E. Liepins, and A. S. Kunnathur (1986). Optimal imputation of erroneous data: categorical data, general edits. *Operations Research* 34(5), pp. 744–751.

Genesereth, M. R. (1984). The use of design descriptions in automated diagnosis. *Artificial Intelligence* 24(1–3), pp. 411–436.

Glover, F. (1988). TABU Search. Technical Report CAAI Report 88–3, Graduate School of Business, University of Colorado, Boulder, Colo.

Goldberg, D. E. (1989). *Genetic Algorithms in Search, Optimization, and Machine Learning*. Reading, Mass.: Addison-Wesley.

Jacobs, R. A. (1988). Increasing rates of convergence through learning rate adaptation. *Neural Networks* 1(4), pp. 295–308.

Josephson, J., Chandrasekaran, J. Smith, and M. Tanner (1987). A mechanism for forming composite explanatory hypotheses. *IEEE Transactions on Systems, Man, and Cybernetics* SMC–17(3), pp. 445–454.

de Kleer, J. de, and B. C. Williams (1987). Diagnosing multiple faults. *Artificial Intelligence* 32(1), pp. 97–130.

Liepins, G. E., M. R. Hilliard, J. Richardson, and M. Palmer (1990). Genetic algorithm applications to set covering and traveling salesman problems. In Brown (ed.), *OR/AI: The Integration of Problem Solving Strategies*.

Liepins, G. E., and D. J. Pack (in press). A simulation study of selected methods for dealing with survey errors. In Liepins and Uppuluri (eds.), *Data Quality Control: Theory and Pragmatics*. Marcel Dekker.

Michalski, R. S. (1987). How to learn imprecise concepts: a method for employing a two-tiered knowledge representation in learning. *Proceedings of the Fourth International Workshop on Machine Learning*. Irvine, Calif.: Morgan Kaufmann, pp. 50–58.

Mooney, R., J. Shavlik, G. Towell, and A. Gove (1989). An experimental comparison of symbolic and connectionist learning algorithms. *Eleventh International*

Joint Conference on Artificial Intelligence. San Mateo, Calif.: Morgan Kaufmann Publisher, pp. 775–780.

Pearl, J. (1984). *Heuristics.* Reading, Mass.: Addison-Wesley.

Peng, Y., and J. A. Reggia (1987a). A probabilistic causal model for diagnostic problem solving. Part I: integrating symbolic causal inference with numeric probabilistic inference. *IEEE Transactions on Systems, Man, and Cybernetics,* SMC–17(2), pp. 146–162.

Peng, Y., and J. A. Reggia (1987b). A probabilistic causal model for diagnostic problem solving. Part II: diagnostic strategy. *IEEE Transactions on Systems, Man, and Cybernetics,* SMC–17(3), pp. 395–406.

Peng, Y., and J. A. Reggia (1989). A connectionist model for diagnostic problem solving. *IEEE Transactions on Systems, Man, and Cybernetics,* SCM–19(2), pp. 285–298.

Potter, W. D., J. A. Miller, and O. R. Weyrich (in press, a). A comparison of methods for diagnostic decision making. *Expert Systems with Applications: An International Journal.*

Potter, W. D., B. E. Tonn, M. R. Hilliard, G. E. Liepins, R. T. Goeltz, and S. L. Purucker (in press, b). Diagnosis, parsimony, and genetic algorithms. *Third International Conference on Industrial & Engineering Applications of Artificial Intelligence and Expert Systems.*

Purucker, S. L., B. E. Tonn, R. T. Goeltz, T. P. Wiggen, K. M. Hemmelman, S. F. Borgs, and R. D. Rasmussen (1989). Design and operation of the communication alarm processor expert system. *Proceedings of the Power Systems and Expert Systems Conference,* Seattle, Wash.

Quinlan, J. R., Induction of decision trees. *Machine Learning* 1, pp. 81–106.

Reggia, J. A., D. Nau, and P. Wang (1983). Diagnostic expert systems based on a set covering model. *International Journal of Man-Machine Studies* 19(5), pp. 437–460.

Reiter, R. (1987). A theory of diagnosis from first principles. *Artificial Intelligence* 32(1), pp. 57–95.

Richardson, J. T., M. R. Palmer, G. E. Liepins, and M. R. Hilliard (1989). Some guidelines for genetic algorithms with penalty functions. In Schaffer, (ed.), *Genetic Algorithms.* San Mateo, Calif.: Morgan Kaufmann, Publisher, pp. 191–197.

Valiant, L. G. (1984). A theory of the learnable. *Communications of the ACM* 27, pp. 1134–1142.

Weyrich, O. R., W. D. Potter, and J. A. Miller (in press). A simulation study of heuristic techniques for diagnostic decision making. *Eastern Multiconference of the Society for Computer Simulation.*

18

A Genetic Algorithm for Conformational Analysis of DNA

C. B. Lucasius, M. J. J. Blommers, L. M. C. Buydens, and G. Kateman

SUMMARY

Conformational analysis of aqueous deoxyribonucleic acid (DNA) molecules is a complex chemical problem that is enjoying increasing scientific interest and is an important tool for some applied sciences. Among techniques for conformational analysis of macromolecules in general, restrained molecular dynamics and distance geometry are, thus far, most widely used. However, both techniques also have some important shortcomings which have been ascribed to problem complexity, but can be circumvented by employing performance-based methods, especially nonlinear adaptive techniques.

In this chapter, the potential of genetic algorithms for conformational analysis of aqueous DNA is discussed. Compared to the "traditional" conformational techniques mentioned, this approach appears to exhibit significantly better sampling of conformational space, leads to conformations unbiased by the expert chemist's intuition, and, as a rule, manages to satisfy all experimental constraints imposed. The "genes" manipulated by our genetic algorithm, called DENISE, are the principal conformational parameters of the given DNA molecule. Trial DNA conformations, maintained in a population, compete for a best fit between theoretical data that can be derived from them and experimental nuclear magnetic resonance data. In addition, DENISE can cope with some noise and internal conflicts in its experimental input data. DENISE's major problem is the presence of strong correlations between conformational parameters in the backbone of DNA; to deal with this problem a nonintegral optimization strategy appears to be imperative. This implies that separate genetic algorithms are used for distinct molecular fragments, which are later combined with larger fragments for which refinement genetic algorithms are used. Apart from this recursive strategy, it appears that additional customizations in evaluation, representation, coding, and control parameter settings are indispensable to fit the problem domain.

INTRODUCTION

Molecular structure provides an efficient means of storing biological data and instructions. The complexity of a molecular structure is inextricably intertwined with its information content. In nature, several kinds of macromolecules play a role as algorithms intended for specific biological tasks. The most fundamental among these macromolecules is deoxyribonucleic acid (DNA) (Saenger 1984), which is the hereditary material of all living systems presently known. Because DNA is a flexible molecule and because it consists of a variable number of units, it can adopt an almost inexhaustible number of structural modes. This accounts for the wide variety of DNA-related phenomena observed by scientists.

In order to understand the processes encoded by DNA, three logical steps must be followed:

1. Experimental data are extracted and collected from a representative sample of the DNA species believed to be responsible for the phenomena observed. This is done by means of laboratory analyses.

2. These data are interpreted in order to obtain a molecular structure. If such an analysis is aimed at the atomic level, it is called a *conformational analysis* (CA).

3. To explain the phenomena observed, either a new structural model is designed or an already existing structural model is adjusted.

Errors may occur in each of these three stages and have to be taken into account.

In practice, the second step appears to be the most cumbersome. Conducting a CA of a large macromolecule by hand may occupy an expert conformational chemist for several months (sometimes even years). The wider availability of increasingly faster and cheaper computers during the past decade has not led to algorithms that can fully take over human interpretation of data measured for conformational purposes. Yet, automating and increasing the reliability of CA procedures are not only very important from a scientific point of view, but for some industrially applied sciences, for example biotechnology,[1] reliable conformational techniques are useful tools. Many CA strategies have already been explored and, to now, *restrained molecular dynamics* and *distance geometry* are most widely used. However, when we deal with complex macromolecules, the outcome of these techniques is not always satisfying (van de Ven and Hilbers 1988). For this reason, recently, conformational chemists have diligently been looking for alternative techniques; it appears that there is increasing consensus with regard to the practical usefulness and potential of conformational techniques based on nonlinear adaptation.[2] Such techniques may lead to more reliable structure elucidation and thus shed new light on the intriguing and important properties of DNA systems that are not yet entirely understood.

This chapter outlines a CA method based on a genetic algorithm we have called DENISE (DNA Evolutionary NOE[3] Interpretation for Structure Elucidation). This method circumvents most of the burdens posed by the "traditional" techniques just mentioned. In the past year of its existence DENISE has matured to the point that CAs of DNA consisting of up to six nucleotides can be performed in a few days on an average mainframe computer.

DNA structure

Biochemists may wish merely to browse this section.

In order to obtain structural information concerning DNA as it exists in living forms, its natural (cellular) environment must be approximated in a laboratory environment by gathering experimental data from samples in

---

[1] Biotechnology has already proved its usefulness in the drug industry, medical practice, agriculture, cattlebreeding, waste processing, and the production of electricity. It has obviously become an important part of our daily lives.

[2] Adaptive algorithms based on nonlinear dynamics are increasingly welcomed and are gaining ground in other chemical areas too (Borman 1989; Lucasius and Kateman 1989; Smits et al. 1990).

[3] NOE denotes the kind of experimental input data used by the genetic algorithm and is discussed later in this chapter.

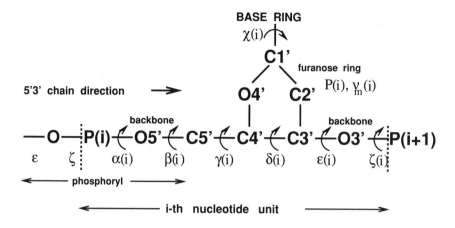

Figure 18.1:   Schematic drawing of one (i^{th}) nucleotide unit, emphasizing back-bone and furanose (sugar) ring. Three-dimensional impression is suppressed: in reality the furanose ring is puckered and bonding angles in the backbone are approximately 110°. Atoms are indexed according to international conventions. Hydrogen (H) atoms are not drawn in order to emphasize the principal conformational parameters, which are: the torsion angles α, β, γ, ϵ and ζ in the backbone, the torsion angle χ along the bond connecting furanose ring and base ring system, and the furanose ring parameters ν_m (pucker amplitude) and P (pseudorotational phase angle).

aqueous solution. Aqueous DNA (henceforth simply referred to as DNA) is a flexible (nonrigid) macromolecule that is distributed among several more or less stable conformations. It is built up from polymerization (chemical chaining) of nucleotides, which accounts for its variable size. Nucleotides are molecules of which four species exist: thymidine (T), guanosine (G), cytidine (C) and adenosine (A). These are the four "letters" that comprise the alphabet used by nature to encode hereditary information. Three functional groups can be distinguished in each nucleotide: *backbone, furanose* (sugar) ring, and *base* (Figure 18.1). The base is a molecular ring system of which four types exist. It uniquely determines the identity of the nucleotide (T, G, C, or A) to which it belongs. Figure 18.2 illustrates some molecular detail of the four base ring systems, which bear names closely related to the nucleotide names mentioned above: thymine, guanine, cytosine, and adenine, respectively. The chemical bonding scheme (atomic connection table) of each nucleotide follows from earlier research conducted by others and this is used as a priori knowledge for DENISE.

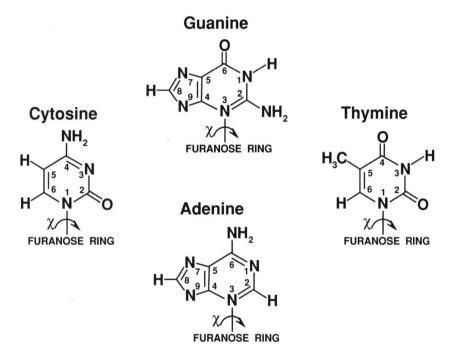

Figure 18.2:   The four-base ring systems of DNA and standard indexing of their atoms.

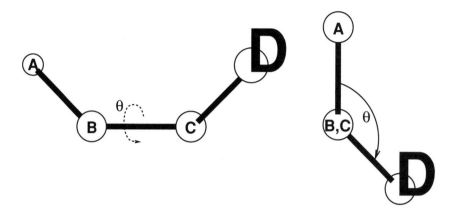

Figure 18.3: Pictorial definition of a torsion angle (indicated as θ).

A distinction is made between the concepts *conformation, structure,* and *sequence.* By definition, a *conformation* specifies exact three-dimensional atomic positions, whereas a *structure* reflects a weighted global average (mixture) of all distinct conformations that exist for a given DNA sequence. A DNA *sequence* is said to be specified when the identity and order of its constituent nucleotides are known. DNA sequence determination appears to pose a comparatively low analytical threshold in laboratory practice. Consequently, CA techniques (including DENISE) usually take a known DNA sequence as input.

The *conformational space* of a DNA molecule is spanned by its conformational parameters (internal degrees of freedom). Each point in this multidimensional geometrical space uniquely represents a conceivable conformation. This vast (but always finite) space is the permutational or theoretical conformational space. Ideally, a conformational problem solver prunes the theoretical conformational space down to a tiny area rather than to a single conformation—the feasible conformational space—which represents the *set* of preferential conformations that collectively provide a probabilistic picture of DNA structure. In this respect, the use of a population-based CA technique is advantageous, since a converged population might reflect an estimate of such set.

There are three elementary types of conformational parameters: *bonding distance* (length of a chemical bond), *bonding angle* (angle between two adjacent chemical bonds), and *torsion angles* along single chemical bonds (Figure 18.3). However, some of these parameters are far less important than the others and can therefore be safely discarded from the genetic algorithm. For instance, reliable estimates of bonding distances and bonding angles in aqueous DNA are obtained from X-ray diffraction data collected from DNA single-crystals, and these may be considered to be reasonably constant. Hence, only torsion angles are taken into account to span the conformational space of DNA.

Some further reductions in conformational space are possible. First, torsion angles along chemical bonds in the base ring systems are considered reasonably constant, because the nature of the chemical bonds is such that this nucleotide section is always essentially flat.[4] Despite its rigidity, a base ring system as a whole possesses all freedom, in principle, to "wiggle" along the single chemical bond that connects it to a furanose ring. The torsion angle responsible for this is called χ (Figures 18.1 and 18.2). Secondly, the so-called endocyclic torsion angles along the chemical bonds in furanose rings are sterically constrained because the furanose ring needs to be closed (Figure 18.1). For this reason these torsion angles can be fully described by two internal rotational parameters only: the *pucker amplitude*, ν_m, and the *pseudorotational phase angle*, P. The endocyclic torsion angles are then given by:

$$\nu_j = \nu_m \cos\left[P + \frac{4\pi(j-2)}{5}\right]$$

where $j = 0, 1, 2, 3, 4$, for ν_j along chemical bonds $O_{4'}-C_{1'}$, $C_{1'}-C_{2'}$, $C_{2'}-C_{3'}$, $C_{3'}-C_{4'}$ and $C_{4'}-O_{4'}$, respectively (Figure 18.1).

In summary, the functional groups in DNA can be listed in order of decreasing internal rotational freedom: backbone, furanose ring, and the flat base ring system. There are eight principal conformational parameters per nucleotide unit—χ, ν_m, P, α, β, γ, ϵ and ζ (Figures 18.1 and 18.2)—which are all constrained to a finite range. The torsion angles α, β, γ, ϵ, ζ and χ, can, in principle, take all values between $0°$ and $360°$; δ is part of both the backbone and furanose ring (where it is called ν_3) and is therefore redundant. Ranges for P and ν_m are derived by applying techniques beyond the scope of this chapter[5]; typical ranges are $150.0—225.0°$ and $25.0—50.0°$, respectively. A conformation of a given DNA sequence consisting of N nucleotides is considered fully determined when its $8N$ conformational parameters are found.[6]

---

[4]The electrons of the bonds are delocalized into a π-system, which is flat.

[5]2D-COSY: 2-dimensional correlation NMR spectrometry

[6] Because of chain-end effects with both mathematical and experimental causes, not all conformational parameters are defined. For instance, since four atoms are required

A DNA sequence is determined not only by the order of its constituent nucleotides, but also by the orientation of the strand (which is either 3'5' or 5'3' (see Figure 18.1). The most common structure of DNA is the so-called antiparallel double helical structure, first discovered by Watson and Crick (1953). This duplex structure, often referred to as *regular* or *Watson— Crick* DNA, consists of two antiparallel DNA strands (3'5' and 5'3') that are paired by virtue of the formation of so-called hydrogen bonds[7] between the basic parts of nucleotides, where T pairs only with A, and G only with C. Several less common DNA structures—aberrant (nonregular) DNA— have also been reported. Both regular and aberrant natural DNA structures possess important biological functionality, related to their chemical structure.

One type of natural DNA complexes currently enjoying widespread attention in the field of biomolecular conformational research belongs to the class of DNA *hairpins* (Blommers et al. 1989; Blommers 1990; van de Ven 1988). These complexes are formed in partially self-complementary sequences of nucleotides. A typical example of a DNA hairpin is illustrated in Figure 18.4, where a (double-stranded) *stem* and a (single-stranded) *loop* can be distinguished. This figure schematically depicts a typical hairpin, which serves as a test case to investigate DENISE's performance (see "Partial Conformational Analysis of a DNA Hairpin"). The stem approximates the Watson—Crick structure, which is already fairly accurately known. The most important part of the hairpin is its loop. Hairpins are well conserved during evolution and therefore may have important biological functions. For instance, hairpins are formed at the origin of replication of bacteriophages (Baas and Jansz, 1988). Hairpins in the DNA genome are believed to be involved in complex interactions and can be recognized by other biological molecules, for example, proteins. Similar unique structural properties are also encountered or expected in even larger DNA complexes that have been observed, such as pseudoknots (Pleij et al. 1985) and compound hairpins like cloverleafs. Aberrant DNA structures such as hairpins, pseudoknots, and cloverleafs are especially in need of improved CA techniques.

---

to define a torsion angle along a chemical bond (Figure 18.3), both chemical bonds comprising the ends of the backbone chain have mathematically undefined torsion angles. Also, at chain-ends there are usually less experimental constraints, and this may cause a few more torsion angles to be undefined as well.

[7]A hydrogen bond is a special kind of chemical bond that is relatively weak.

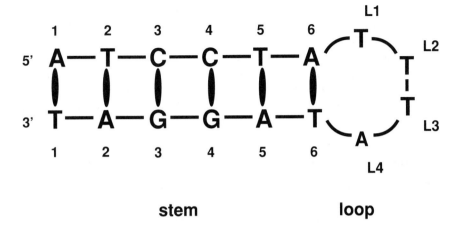

Figure 18.4: Schematical drawing of the d(ATCCTA-TTTA-TAGGAT) hairpin emphasizing stem and loop section. By convention, the four nucleotides in the loop section are also indicated by L_i $(i = 1, 2, 3, 4)$.

NOE Spectra as Experimental Input Data

NMR experts may wish merely to browse this section.

If the H-atoms omitted in Figure 18.1 are included, the number of chemical bonds depicted at each C-atom is completed to 4. H-atoms in the base sections of DNA are illustrated in Figure 18.2. Since DNA is apparently abundant in H-atoms and since nuclear magnetic resonance (NMR) spectrometry is sensitive to the magnetic properties of among others H-atoms, it makes sense to collect experimental input data for CA of DNA by means of this instrumental technique.

Although several kinds of NMR spectrometry are available, this chapter is concerned only with two-dimensional (2D) nuclear Overhauser enhancement (NOE) spectrometry. Basically, the information contained in 2D-NOE spectra "pins down" relative H-positions in the molecule, leading to geometrical constraints (Macura and Ernst 1980). Ideally, then, a CA proceeds in two steps:

Ideal Conformational Analysis Strategy:

Experimental 2D-NOE spectrum $\Longrightarrow$ Set of relative H–H distances $\Longrightarrow$ DNA conformation.

If this strategy were feasible in practice, 2D-NOE spectrometry would be very useful for conducting routine CA of biological macromolecules in general, for it is a very informative instrumental technique. However, the problem is that neither of the steps is mathematically well defined. On the other hand, if these steps are reversed, straightforward mathematical relationships apparently do exist. This implies that 2D-NOE spectrometry can in practice be used in combination with an iterative performance-based strategy:

Iteration in a Performance-based Conformational Analysis Strategy:

Proposed DNA conformation $\implies$ Set of relative H–H distances $\implies$
Theoretical 2D-NOE spectrum:
Performance = goodness of fit with experimental 2D-NOE spectrum.

This strategy seems tailor-made for implementation in a genetic algorithm and, indeed, comprises the backbone of DENISE. Another reason why a genetic algorithm is believed to be useful is that a converged population might reflect an estimate of the *family* of conformations defined by NOE-imposed constraints. Figure 18.5 illustrates the principle of 2D-NOE spectrometry. Through-space magnetic interactions between pairs of H-atoms lead to off-diagonal peaks that contain distance information. More precisely, distance is roughly inversely proportional to the sixth power of observed peak intensity (Macura and Ernst 1980). Experimental 2D-NOE spectra are processed by a peak-find procedure that calculates the intensity of each peak and identifies the two H-atoms belonging to it. The *NOE table* thus obtained lists an intensity for each H—H combination and as such is suitable as experimental input data for CA. Albeit a set of H—H distances can not be derived in a straightforward way from a NOE table, the latter may still be processed further to obtain an *H-H distance constraints table*. It is similar in form, but it lists a distance interval for each H—H combination instead.[8] The reason why distances cannot be accurately derived from NOE intensities is that H—H pairs should not be treated as isolated systems, that is, the presence of other H-atoms is "felt," leading to contributions to primary H—H interactions. For the purposes of this chapter, it suffices to denote such secondary contributions as

---

[8]Optionally, the distance constraints table may be extended to contain distance constraints of non-H—H atom pairs too, if available.

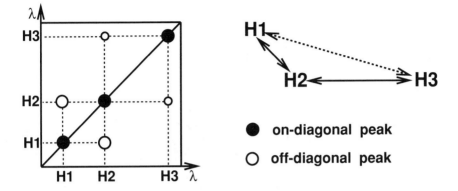

Figure 18.5: Schematic top view of a 2D-NOE spectrum for a molecule consisting of three hydrogen (H) atoms. Peaks are plotted as a function of two wave numbers (ω). A short-range interaction (H1—H2) leads to a peak with higher intensity as compared to a long-range interaction (H2—H3). H-atoms may be too far apart to give any measurable interaction at all (H1—H3).

H$\cdots$(H—H) interactions.[9] When H$\cdots$(H—H) interactions are taken into account, corrections and confidence intervals can be estimated for the H—H distances obtained in the hypothetical case in which these interactions are not present (Macura and Ernst 1980). Despite their inherent error, distance constraints tables are useful for enabling significant reduction in the overall computational load of the genetic algorithm (see "Cascaded Hybrid Evaluation").

CONFORMATIONAL ANALYSIS IN PERSPECTIVE

Both restrained molecular dynamics and distance geometry use, as *on-line* input, atomic distance constraints derived from experimental NOE tables. Experimental NOE tables, on the other hand, are not used on-line but may be used as an *off-line*, that is, a posteriori, means to verify whether conformations found by either of these techniques are reasonable. Such verifications often appear to be nonaffirmative for complex macromolecules, which is why the scope of these techniques is limited in practice. Restrained molecular dynamics is based on a force field model, whereas distance geom-

---

[9] *Spin diffusion* in expert terminology.

etry uses a set of geometrical rules. Neither method can be applied without certain radical simplifications; this is regarded as the primary cause for poor results when complex macromolecules are dealt with. Distance geometry generates structures from geometrical constraints in an *ab initio* way. The method has the advantage of essentially generating unbiased conformations because no knowledge of the structure is used as input. Nevertheless, distance geometry exhibits poor sampling properties, as strongly unwinded structures are commonly obtained (Metzler et al. 1989). Restrained molecular dynamics, also, exhibit poor sampling properties, because a starting structure is needed as input. This implies that results may depend on the scientist's intuition because sampling of conformational space will be concentrated around this structure. Restrained molecular dynamics and distance geometry belong to a problem-solving category in which strategies based on mathematical or heuristic rules are employed. Increasing problem complexity requires increasing rule complexity, but explication of the latter is required for this and appears to be limited by practical feasibility. A sensible alternative to rule-based problem solving is *performance-based* problem solving. For a broad class of complex problems a performance measure can often, counterintuitively, be formulated in a surprisingly plausible and straightforward fashion. Conformational chemists are increasingly realizing that complex structure elucidation problems require approaches that are not based on dramatic mathematical assumptions and approximations. Recent scientific papers in the field of CA clearly indicate a paradigm shift in favor of performance-based problem solving. Our application is one among several that follow this line. Next, we present some examples and give a brief comparison with the genetic algorithm.

Lipton and Still (1988) report a CA method based on exhaustive tree (grid) search, combined with a set of criteria designed to eliminate chemically unreasonable conformations and thus speed up search. An important disadvantage of exhaustive search, as compared to an approach based on a genetic algorithm, is that it has no "memory" to exploit the intermediary knowledge obtained. In practice, then, it is limited to relatively small molecules.

Wilson et al. (1988) propose simulated annealing, which is based on stochastic search (sampling), as a promising method of performing global energy minimization of flexible organic molecules. Good sampling properties are reported for simulated annealing in various other applications too, but, as a rule, genetic algorithms have better sampling properties owing to implicit parallelism (Holland 1975), that is, to selectionist interactions between population members. Systems that exhibit collective behavior based on intensive interactions are highly nonlinear. This renders the genetic algorithm an important exception among performance-based problem solvers, because nonlinear problem spaces, for example, conformational spaces of complex macromolecules (see the subsection "Phosphoryl-related Correlations Between Conformational Parameters"), tend to insist on nonlinear problem solvers. This is also recognized by Wu and Freeman (1989), who review the potential of Darwinism in general as applied to magnetic resonance spectrometry. Another non-linear problem solving concept with much potential for CA of complex macromolecules is embodied in neural networks, in which collective behavior is guided by adaptation and self-organization. Apart from our implementation of a genetic algorithm for study of DNA folding (DENISE), Qian and Sejnowski (1988) investigate protein folding based on a neural network that learns from examples. This approach requires that the expert chemist practice great care in selecting the set of examples intended to be representative for learning. By contrast, DENISE essentially utilizes instrumental input data (see the subsection "NOE Spectra as Experimental Input Data" above) and is probably less sensitive to human bias.

REPRESENTATION, CODING AND CONTROL

By proposing values for the conformational parameters of the DNA sequence subject to CA, a trial conformation is obtained. DENISE maintains a population of such trial conformations, each of which is represented as a string ("chromosome") of values ("genes") for the conformational parameters. DENISE manipulates these genes in its attempt to find optimal conformations of the DNA molecule. A distinction is made between the concepts *representation* and *coding*. Representation involves the meaning of genes (as just indicated) and their order on a string, whereas coding involves their form of appearance (e.g., binary-coded, real-coded). (The importance of gene order is discussed later in this chapter.) As far as gene coding is concerned, in the first instance, real coding might be construed as most useful; after all, the rotational parameters to be obtained are all real. This approach is described in Lucasius and Kateman 1989, and requires modified genetic operators for optimal performance.[10] In the present chapter, however, we follow the approach encountered in the mainstream literature on genetic algorithms, which is binary coding. A binary-coded

---

[10]For instance, mutation is subdivided into jump mutation and step mutation, which correspond to Davis's big creep and little creep mutation, respectively (Davis 1989). Also, mutation rates are higher than usual.

(encoded) gene is a bitfield that consists of a fixed number of bits, which needs to be interpreted (decoded) to obtain the real number it represents. We can take advantage of the fact that all genes in the present application represent real numbers with a *finite range*. The unsigned integer value corresponding to a bitfield can therefore be used as a discrete quantifier for the real parameter range it applies to. More precisely, if L = gene lowbound (real), H = gene highbound (real), B = number of bits in gene (bitfield), and E = encoded gene value (unsigned integer of B bits), then the decoded gene value D (real) is given by

$$D = L + (H - L)\frac{E}{2^B}$$

This equation states that parameter values are the decoded (i.e., real-coded) incarnations of the binary-coded genes of the GA. Obviously, the bitfield size B dictates the resolution that can be attained in determining the decoded parameter value and should therefore be set as a control parameter. Instead of treating a gene bitfield as a binary integer, it may be treated as a so-called Gray integer (Caruana and Schaffer 1988). When Gray integers are first transformed to binary integers, the above decoding equation can be used again, substituting E by the transformed integer. Caruana and Schaffer (1988) have empirically found that Gray coding is often better than binary coding for genetic algorithms that involve numerical parameters.[11] So far, our results suggest that Gray coding is not only superior but even indispensable for CA of DNA sequences consisting of more than two nucleotides (see "Phosphoryl-related Correlations Between Conformational Parameters").

Since genetic operators (selection, recombination, crossover, mutation, etc.) are not "aware" of the meaning of the bit strings that compose a population, a toolbox of genetic operators—the *genetic engine*, that is, the GA's "core" that drives a given application—can be separated from the problem domain, allowing maximum flexibility in program development. Hence, changes in one of the constituent modules imply a minimum need for changes in the other to preserve mutual consistency. DENISE is configured in this modular fashion. The domain-dependent module is "plugged" into the genetic engine module by establishing pointer references. The architecture of the genetic engine complies with "standard" schemes for genetic algorithms reported by others. See, for instance, Goldberg's comprehensive textbook (Goldberg 1989a) or Holland's seminal work (Holland 1975).

---

[11]More generally, this is true for *ordinal* parameters, i.e., for parameters with values that obey some ordering criterion.

DENISE is written in the C language and runs on all computers for which C compilers compatible with the ANSI standard are available, for example, IBM-PC's and SUN computers. It takes the DNA sequence of interest as a character string argument, for example, ATTTAT. The setting of control parameters is arranged in two files, one for domain-independent parameters (e.g., mutation rate, crossover rate, coding fashion) and the other for domain-dependent parameters (e.g. the NOE evaluation weight factor in "Cascaded Hybrid Evaluation" and the refinement sensitivity in "Recursive Pairwise Refinement"). Intuitive prediction of the impact of customizations in control is very tricky when we are dealing with nonlinear systems (Goldberg 1989b). Therefore, we have chosen to set the domain-independent control parameters to "regular" values, that is, to rely on suggestions of genetic algorithm theoreticians and on estimates made by meta-genetic algorithms (Grefenstette 1986).

CREATION OF TRIAL CONFORMATIONS

Once the conformational parameters on a given bit string are decoded, they are processed to obtain a trial conformation, which implies that the spatial positions of important atoms (among which is H) are calculated with respect to some arbitrary three-dimensional Cartesian frame $\mathcal{F}$. The atomic positions cannot be derived all at once from the conformational parameter values: an iterative "bookkeeping" procedure is required, which we call *molecular parsing*. Each parsing step can be described formally by an algebraic frame transformation with three components: a translation (shift) of $\mathcal{F}$'s origin along a chemical bond by its bonding distance; a rotation of $\mathcal{F}$ along one of its axes by a molecular bonding angle; and a rotation of $\mathcal{F}$ along another of its axes by a molecular torsion angle. In the parsing procedure, $\mathcal{F}$ moves stepwise along the molecule—from atom to adjacent atom and so on—introducing a new atom in each parsing step while updating the positions of all earlier introduced atoms with respect to the new frame $\mathcal{F}$ obtained right after this parsing step. This goes on until all atoms in the molecule have been scanned. The goniometrics required for frame transformations in molecular parsing are straightforward and can be found in Hendrickson (1961), for example.

CASCADED HYBRID EVALUATION

Once the relative atomic positions in a proposed conformation are calculated, fitness is allocated to the bit string representing it. DENISE employs a *cascaded hybrid evaluation criterion* for this purpose; that is, the fitness of a chromosome is computed by two different procedures, in sequence. The first contribution is provided by a module which calculates an H—H distance table for each trial conformation that is compared to the H—H distance constraints table derived from an experimental 2D-NOE spectrum. This fitness contribution is proportional to the number of H—H distance constraints satisfied.

Only when all distance constraints are satisfied can the second evaluation module be invoked. Using the so-called Wangsness—Bloch—Redfield relaxation theory, summarized in van de Ven and Hilbers (1985), a *theoretical* NOE table is calculated from the relative H-positions of each trial conformation, and a goodness of fit value is derived from a comparison with the experimental NOE table. This fitness contribution is multiplied by a weight factor—the *NOE evaluation weight factor*—and is then added as bonus fitness to the former contribution. The NOE evaluation weight factor is a domain-dependent control setting parameter, which is set to a value larger than 1 to indicate that NOE tables are, as a rule, more reliable than distance constraints tables.

As opposed to the techniques of restrained molecular dynamics and distance geometry, DENISE uses experimental NOE tables as *on-line* primary input. This is a great advantage because NOE tables are much more informative and reliable than distance constraints tables (the on-line input of restrained molecular dynamics and distance geometry). In DENISE, distance constraints tables are used merely to economize on the overall computational load of the genetic algorithm, which is dominated by NOE evaluations. These evaluations are computationally very intensive as a consequence of the numerous matrix multiplications involved. Such calculations should, therefore, be invoked only when a first guess of the optimal state is obtained by satisfying the distance constraints. For the sake of flexibility, however, DENISE can be used in the nonhybrid evaluation mode too, which implies that only one of the two evaluation units is activated.

DEALING WITH PROBLEM COMPLEXITIES

Molecular Flexibility

DNA's inherent internal flexibility implies that a given DNA sequence manifests itself as a dynamical *mixture* of single conformations, that is, that a 2D-NOE spectrum (cq. NOE table) is in practice a superposition of several hypothetical monoconformational 2D-NOE spectra (cq. NOE tables). This means that, as far as an experimental NOE table pretends to refer to a single DNA conformation—the prominent conformation—*internal conflicts* are present in the NOE table. This problem is a very subtle one, because the intensities listed in the experimental NOE table depend not only on H—H distances but also on conformational concentrations (probabilities). Therefore, a weak NOE intensity might belong to a very probable conformation, and vice versa. In principle, then, there is potential jeopardy that the search will become biased toward relatively unimportant conformations. The performance of a genetic algorithm in this complex situation can be verified only if the probability distribution of DNA conformations pertaining to the DNA sequence of concern is a priori known, but it is practically impossible to derive such a distribution from an experimental NOE table. Conversely, an experimental NOE table can be simulated from a proposed probability distribution of conformations that pertain to a given DNA sequence. This "experimental" NOE table is a weighted sum—weighted in proportion to mentioned probability distribution—of monoconformational NOE tables that are theoretically derived as indicated in the preceding section. Some simulated instrumental noise may be added to approximate laboratory conditions. The usefulness and validity of simulations as just described are discussed later in this chapter.

Simulated experimental NOE tables were created for a mixture of selected DNA conformations, among which one was chosen to be ten times more probable than the others. DENISE was then started with the options Gray coding and sharing activated, to analyze these NOE tables. There was always convergence toward the prominent conformation, and sometimes even separate stable subpopulations emerged for the other conformations too. These observations strongly support the statement that a genetic algorithm itself can carry out the conflict resolution task; that is, it recognizes and separates the monoconformational NOE tables in a fashion loosely similar to Fourier analysis of compound signals. More investigations are needed, however, to show the precise mechanism behind these observations.

H···(H—H) Interactions and Instrumental Noise

H···(H—H) interactions and instrumental noise are the least cumbersome among the complexities faced by DENISE. They may pose some difficulties only when relatively complex DNA molecules are considered. The Wangsness—Bloch—Redfield theory, used to create theoretical NOE tables, is based on some *assumptions* to estimate H···(H—H) contributions to the H—H interactions that it predicts. These assumptions are not too radical and are satisfied quite well in practice (van de Ven 1988).

NMR spectrometers with very good signal-to-noise ratios are widely available in contemporary laboratory practice. Our observations confirm that genetic algorithms perform well in noisy environments. Specifically, we found that DENISE is quite insensitive to the imposition of additional noise to experimental NOE tables, as closely reminiscent conformations were repeatedly obtained in both cases.

Multimodalities in Conformational Space

Any macromolecular conformational space is highly multimodal; that is, it consists of an immense number of optimal regions, ranging from suboptimal to real-optimal. Sub-optima act as traps in which iterative search techniques, among which are genetic algorithms, may get stuck. Although genetic algorithms are known to be fairly insensitive to premature convergence, a very large number of (sub-)optima might frustrate even a genetic algorithm. Such problems are referred to as *GA-hard* (Goldberg 1989a).

This problem may be alleviated by applying *sharing* operators (Deb and Goldberg 1989). The effect of sharing is that stochastically stable subpopulations will collect on (sub-)optima, where the size of each subpopulation is proportional to the average local performance experienced. This dispersal property is very desirable, because the best conformations may be found in near-optima rather than in real optima as a consequence of internal conflicts in experimental input (Section 18) and other problem complexities. Deb and Goldberg (1989) have found that, in general, phenotypical sharing is better able to maintain stable subpopulations on (sub-)optima. As a result, only this sharing mode has been implemented in DENISE.

Phosphoryl-related Correlations Between Conformational Parameters

In the simplest case of multimodality, (sub-)optima are separate points in problem space. More complex multimodalities entail (sub-)optimality *hy-*

perplanes in problem space. This situation occurs when conformational parameters are *correlated*, (i.e., interdependent, coupled), and it often introduces nonlinearities in problem space. Correlations may also be viewed as information sinks because the valid solution set is usually a subset of the optimality hyperplane and so cannot be uniquely determined. The absence of H-atoms in *phosphoryl groups* ($-O_{3'}-P-O_{5'}-$ sections in the DNA backbone, Figure 18.1) leads to strong correlations between the conformational parameters describing these groups (ϵ, ζ, α, β). This is believed to be the principal problem complexity which DENISE must confront. In general, other conformational parameters are also correlated but usually to a significantly lesser or negligible extent. To understand phosphoryl-related correlations between conformational parameters, consider two H-atoms in the DNA molecule, each at opposite ends of a phosphoryl group. Then, several combinations of torsion angle values in the phosphoryl group are conceivable for which the H—H distance is constant. These combinations thus follow a hyperplane. The presence of more H-atoms imposes more restrictions (i.e. reduces the dimensionality of the hyperplane). A reduction to dimensionality 0—this represents a unique conformation—can be attained only if the H-atoms are involved in enough measurable NOE interactions. If this condition is not satisfied, the NOE table lacks essential information to conduct a reliable CA. It is then likely that biochemically nonsensical conformations are found. Even if the H-atoms áre involved in enough measurable NOE interactions to establish a unique conformation, the performance course near this optimum is very deceptive. In fact, it has been *empirically* found that conformations that satisfy practically all experimental constraints are obtained only when the following conditions apply at the same time:

- The sequence or sequence fragment dealt with contains at most one phosphoryl section (Figure 18.1).

- Sharing is active.

- Gray coding is used.

- Strongly correlated conformational parameters are represented as juxtaposed genes (tightly linked genes).

In light of what is known as *GA-deceptivity* (Goldberg 1989a), the fourth condition is perhaps most plausible. Since bit strings are allocated high fitnesses only when the correlated parameters approximate certain values *simultaneously*, juxtaposed representation of correlated genes will minimize the disruptive action of crossover operations.

RECURSIVE PAIRWISE REFINEMENT

Since a genetic algorithm can apparently handle at most one phosphoryl section in the DNA backbone at a time, we introduce a *recursive pairwise refinement* strategy. This strategy implies that separate genetic algorithms are run for distinct molecular fragments (one or more nucleotide units) such that search is focused predominantly on at most one phosphoryl section at a time. At a later stage, these fragments are combined with larger fragments for which refinement genetic algorithms are run. From this point of view, a genetic algorithm will reduce the size of each original conformational parameter range, L, to size l:

$$l = \varepsilon\rho + L(1 - \rho) \qquad 0 \le \rho \le 1$$

where ε is an estimate of the parameter prediction error and ρ is called the *refinement sensitivity*: a setting parameter that controls the sensitivity for range reduction. (Reduction is maximal for $\rho = 1$, whereas for $\rho = 0$ conformational parameter ranges do not reduce at all.)

Upon joining molecular fragments, the reduced conformational parameter ranges become the parameter ranges for the refinement genetic algorithm. Since these estimated parameters evidently need relatively little improvement, the refinement genetic algorithm can focus its attention almost entirely on estimating the conformational parameters of the phosphoryl section that links the fragments. Recursive pairwise refinement can be described in a semiformal way. Let S be the DNA sequence under investigation, $\|S\|$ be the number of nucleotides in S, $S_{\frac{1}{2}}$ be one-half of S (or nearly so) such that it consists of $\frac{\|S\|}{2}$ nucleotides if $\|S\|$ is even and $\frac{\|S\|+1}{2}$ nucleotides if $\|S\|$ is odd, and $\tilde{S}_{\frac{1}{2}}$ be the complementary part of $S_{\frac{1}{2}}$. Then recursive pairwise refinement can be described by an operator $\mathcal{R}$ defined as:

$$\mathcal{R}(S) \longrightarrow \begin{cases} \text{If } \|S\| = 1 \text{ or } \|S\| = 2 & \text{Run genetic algorithm for } S. \\ \\ \text{Otherwise} & \mathcal{R}(S_{\frac{1}{2}}), \ \mathcal{R}(\tilde{S}_{\frac{1}{2}}); \ \text{ then:} \\ & \text{Run genetic algorithm for } S. \end{cases}$$

Figure 18.6 illustrates the principle of recursive pairwise refinement for a CA involving a sequence that consists of six nucleotides. Although the idea behind recursive pairwise refinement is attractive, there are some practical marginalia with regard to upward propagation of errors in the CA tree (Figure 18.6). First, if one of the genetic algorithms in the CA tree produces

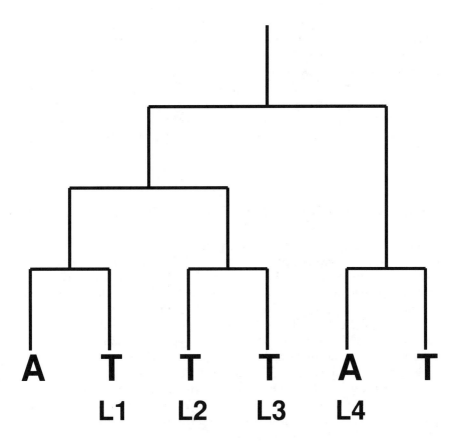

Figure 18.6:   Principle of recursive pairwise refinement illustrated for a conformational analysis of the subsequence A-TTTA-T (A-$L_1L_2L_3L_4$-T) in the hairpin
d(ATCCTA-TTTA-TAGGAT) illustrated in Figure 18.4.

an erroneous estimate of a conformational parameter, it is likely to pass to its refinement genetic algorithm a parameter range that excludes the valid parameter value. Such errors can, in part, be reduced by decreasing the refinement sensitivity ρ, but for larger fragments this causes search to be extended more toward other phosphoryl groups too and this was just mentioned as an insurmountable complication. Second, instead of using the complete experimental NOE table, only those entries that are relevant to the fragment of the macromolecule concerned are used. This introduces some error, because the entries in a NOE table may in fact not be considered mutually independent, that is, H···(H–H) interactions may not be neglected. However, this problem is partially solved by the recursive pairwise refinement strategy itself, because larger parts of the experimental NOE table are used as larger molecular fragments come up for refinement in time.

THE USEFULNESS AND VALIDITY OF SIMULATIONS

Real experimental NOE tables for some complex DNA systems with poorly known structure have been presented to DENISE. From these studies it follows that the principal condition for successful CA—convergence within the level of experimental noise present—is satisfied. However, this condition is *not* sufficient, since the outcome must be biochemically relevant too. This can be verified for those global structural aspects that are already known, but verification of new aspects is obstructed because the most widely used CA techniques—restrained molecular dynamics and distance geometry— are less useful when complex DNA molecules are considered. For this reason (and also because DENISE has not yet reached its final developmental stage) it makes sense to use *simulated* experimental NOE tables for the moment, provided that these are representative of *real* experimental NOE tables.

Tests with a number of structurally well-defined DNA sequences may help affirm the validity of using simulated experimental NOE data. A few conformations typical for a given structure are then selected and theoretical NOE tables are calculated for them as outlined earlier. A simulated experimental NOE table of the mixture is obtained by weighted superposition of the theoretical NOE tables and addition of simulated instrumental noise (see the earlier subsection, "Molecular Flexibility"). Tests of this kind reveal that for both simulated experimental NOE tables and real experimental NOE tables, DENISE finds conformations close to the a priori

	α	β	γ	ϵ	ζ	χ	P	ν_m
A	–	–	55.0	205.0	280.0	222.0	180.0	40.0
T (L_1)	289.0	201.0	50.0	182.0	255.0	229.0	164.0	41.0
T (L_2)	177.0	182.0	55.0	240.0	235.0	217.0	178.0	43.0
	147.1*	30.3*	54.9	240.7	234.7	217.1	177.9	43.0
T (L_3)	299.0	198.0	78.0	253.0	80.0	187.0	198.0	41.0
	299.5	197.4	77.8	170.3*	132.3*	186.9	197.9	41.2
A (L_4)	73.0	197.0	191.0	184.0	243.0	145.0	181.0	41.0
T	303.0	168.0	57.0	–	–	240.0	155.0	37.0

Table 18.1: Target Values of Conformational Parameters for the Principal Conformation in a Simulated Mixture of A-TTTA-T (A-$L_1L_2L_3L_4$-T) Conformations. Values for the L_2L_3 section as predicted by the genetic algorithm (in NOE-only evaluation mode) dedicated to this fragment are also indicated. (Entries marked by *'s are not defined for reasons mentioned in footnote 6.)

known structure. These observations strongly support the validity of relying on simulated data in general.

PARTIAL CONFORMATIONAL ANALYSIS OF A DNA HAIRPIN

To this point our research has focused primarily on partial CA of the A-TTTA-T (A-$L_1L_2L_3L_4$-T) *loop* section in the previously-displayed DNA hairpin d(ATCCTA-TTTA-TAGGAT) (Figure 18.4). Starting from preliminary knowledge about the structure of this loop, an artificial mixture of five more or less reminiscent conformations was created, and one of these was selected as ten times more probable than the others. For this principal conformation the conformational parameter values are shown in Table 18.1. Next, a simulated experimental NOE table was created from the mixture of conformations, as outlined earlier in this chapter. From this NOE table, in turn, an H—H distance constraints table was derived, as explained in "NOE Spectra as Experimental Input Data." Finally, several CAs of the DNA sequence at hand were performed according to a recursive pairwise refinement strategy, as described above. In a number of these CAs, both the distance constraints table and the simulated experimental NOE tables were used (see "Cascaded Hybrid Evaluation"), and in the others only the experimental NOE table was used (henceforth referred to as NOE-only evaluation mode).

The initial conformational parameter ranges were set to $25.0°$—$50.0°$ for ν_m, $150.0°$—$225.0°$ for P, and $0°$—$360°$ for α, β, γ, ϵ, ζ and χ. The gene bitfield size (decoding resolution) and the population size for each genetic algorithm in the CA tree (Figure 18.6) were set to 10 bits and 100 members, respectively. Both Gray coding and (phenotypical) sharing were activated. Domain-independent control parameters were set to "regular" values, the refinement sensitivity was set to 0.8, and for the CAs in cascaded hybrid evaluation mode the NOE evaluation weight factor was set to 3.0.

The behavior of all genetic algorithms in the CA tree appeared to be quite similar. Therefore, it suffices to present in some detail the results of only one of these genetic algorithms, dedicated to one specific molecular fragment (see the following subsection) and merely to summarize the results for the total DNA sequence A-TTTA-T here.

In cascaded hybrid evaluation mode, convergence was typically attained ten times faster than in the NOE-only evaluation mode. Optimization time appeared to depend strongly on the choice of the initial population of conformations (see "Incorporation of Domain-Specific Knowledge"), but the conformations eventually found did not appear to depend strongly on such a choice. The estimation error for predicting the conformational parameters of the prominent conformation was less than $1°$ (as shown in Table 18.1 for the genetic algorithm dedicated to fragment L_2L_3 in A-TTTA-T). All distance constraints and NOE constraints were satisfied within the level of noise present. This finding strongly suggests that a recursive pairwise refinement strategy is robust toward the error introduced by partial use of experimental NOE tables and that this strategy is likely to be suitable for larger DNA sequences too.

Results for L_2L_3 in A-TTTA-T

The results for the genetic algorithm in the CA tree dedicated to molecular fragment L_2L_3, conducted in NOE-only evaluation mode, are illustrated in Figures 18.7, 18.8, 18.9 and Table 18.1. In NOE-only evaluation mode, which is computationally very intensive, typically 0.45 seconds are required on an average mainframe computer to evaluate each L_2L_3 conformation. This corresponds to 50 hours for 4000 generations of a population consisting of 100 conformations. Although $4 \cdot 10^5$ evaluations seem a myriad for the purpose of obtaining a DNA conformation that satisfies experimental constraints, from a simple calculation it follows that a comparable CA conducted by means of grid search would require, on the average, about 10^{10} times this number of evaluations to approach valid conformational

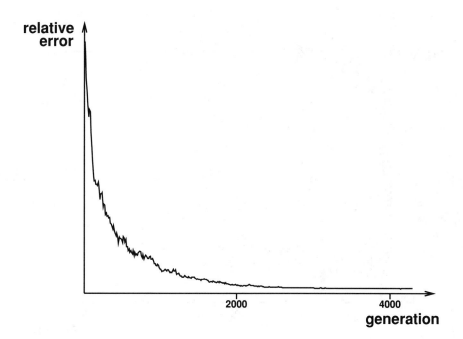

Figure 18.7: Error of fit between the L_2L_3-relevant part of a simulated experimental NOE table of the hairpin loop A-TTTA-T (A-$L_1L_2L_3L_4$-T) and the theoretical NOE table of the best conformation at each generation.

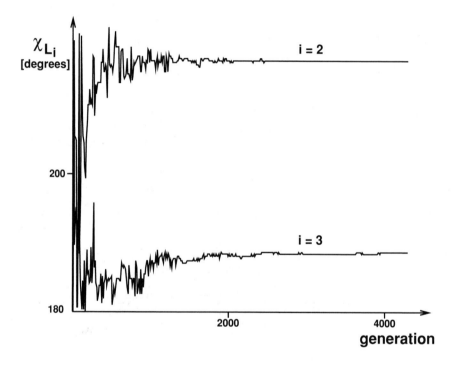

Figure 18.8: Estimated values of torsion angles χ in L_2L_3 section of hairpin loop A-TTTA-T (A-$L_1L_2L_3L_4$-T) of the best conformation at each generation (NOE-only evaluation mode).

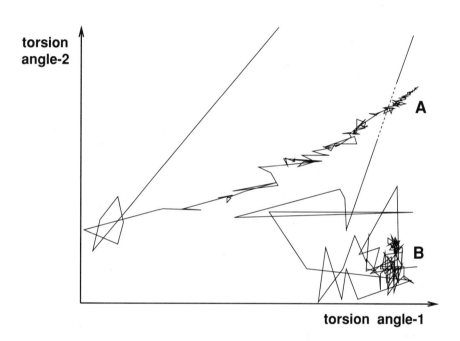

Figure 18.9: Trajectory in (rescaled) projected conformational space (NOE-only evaluation mode) of the best conformation of the L_2L_3 section in the hairpin loop A-TTTA-T (A-$L_1L_2L_3L_4$-T), emphasizing the extent of correlations between conformational parameters. A: in $\zeta_{L_2}-\beta_{L_3}$ plane (strongly correlated torsion angles). B: in $\chi_{L_2}-\chi_{L_3}$ plane (weakly correlated torsion angles).

parameter values with an accuracy of only 10 percent of the respective conformational parameter ranges.

INCORPORATION OF DOMAIN-SPECIFIC KNOWL-EDGE

Optimization time may be considerably reduced when *reliable* domain-specific knowledge is taken into account. For instance, interatomic distance constraints can be derived from knowledge about base pairing. The hydrogen bonds involved in base pairing typically range from 1.5 Å to 2.0 Å. This fact, combined with the rigidity of the base sections, implies that distances between atoms in a base pair are confined to a very narrow range.

The ranges of conformational parameters (χ, ν_m, P, α, β, γ, ϵ and ζ in Figures 18.1 and 18.2) can often be reduced in a reliable way upon expert inspection of 2D-NOE spectra, which may dramatically prune conformational space and thus reduce optimization time considerably too. In general, biased initial populations comprise another sensible means to incorporate domain-specific knowledge (Grefenstette 1987), and CA of DNA by a genetic algorithm appears to lend itself well for this purpose. For instance, for a target DNA sequence consisting of a hairpin loop and stem (e.g., Figure 18.4), we may choose to initialize a population nonrandomly in such a way that the structure of the stem already approximates regular DNA.

CONCLUSIONS AND FUTURE PERSPECTIVES

Genetic algorithms are suitable for large-scale chemical optimization problems such as conformational analysis of aqueous DNA molecules. The expert chemist's knowledge can be incorporated in several ways to reduce computational load, and the impact of such a bias can be adjusted down to the point where only instrumental data are relied on. As compared to restrained molecular dynamics and distance geometry, the results so far suggest that the sampling properties of a genetic algorithm for conformational analysis are superior, since biochemically reasonable conformations are obtained that satisfy experimental constraints within the level of noise present.

Cooperation with local optimization methods has also been experimented with, e.g. energy minimization. This is a useful means for refining the "best guess" (outcome) obtained by the genetic algorithm approach. In its final developmental stage, DENISE is intended to be provided with a user interface so that it can be employed for routine conformational analyses, provided that the optimization times of several days, weeks, or even months (depending on the size of the DNA sequence involved) present no objection.

Genetic algorithm theoreticians are encouraged to contrive alternative representations and codings for DNA conformations, so that the complexities caused by correlations between the conformational parameters that describe the phosphoryl sections in the backbone of DNA can be further reduced.

ACKNOWLEDGMENTS

Prof. Dr. C. W. Hilbers and Dr. F. J. M. van de Ven from the Laboratory for Biophysical Chemistry (Catholic University of Nijmegen) are gratefully acknowledged for their fruitful discussions and the NOE datasets they granted. A. H. J. M. van Aert and S. Werten are mentioned for their active involvement in the project. This research is supported by the Dutch Foundation for Chemical Research (SON) with financial aid from the Dutch Organization for Scientific Research (NWO).

REFERENCES

Baas, P. D. and H. S. Jansz (1988). Single-stranded DNA phage origins. *Current Topics in Microbiology and Immunology* 136, 31. Berlin-Heidelberg: Springer-Verlag.

Blommers, M. J. J., J. A. L. I. Walters, C. A. G. Haasnoot, J. M. A. Aelen, G. A. van der Marel, J. H. van Boom, and C. W. Hilbers (1989). Effects of base sequence on the loopfolding in DNA hairpins. *Biochemistry* 28, 7491.

Blommer, M. J. J. (1990). *Aspects of Loop Folding in DNA Hairpins*. Ph.D. Thesis.

Bohr, H. and S. Brunak (1989). A travelling salesman approach to protein conformation. *Complex Systems* 3:9.

Borman, S. (1989). Neural network applications in chemistry begin to appear. *Computers and Engineering News* 24.

Caruana, R. A. and J. D. Schaffer (1988). Representation and hidden bias: Gray vs. binary coding for genetic algorithms. In J. Laird (ed.), *Proceedings of the Fifth International Conference on Machine Learning.* San Mateo, Calif.: Morgan Kaufmann.

Davis, L. (1989). Adapting operator probabilities in genetic algorithms. In J. D. Schaffer (ed.), *Third International Conference on Genetic Algorithms.* San Mateo, Calif.: Morgan Kaufmann, p. 61.

Deb, K. and D. E. Goldberg (1989). An investigation of niche and species formation in genetic function optimization. In J. D. Schaffer (ed.), *Third International Conference on Genetic Algorithms.* San Mateo, Calif.: Morgan Kaufmann, p. 42.

Goldberg, D. E. (1989a). *Genetic Algorithms in Search, Optimization and Machine Learning.* Reading, Mass.: Addison-Wesley.

Goldberg, D. E. (1989b). Zen and the art of genetic algorithms. In J. D. Schaffer (ed.), *Third International Conference on Genetic Algorithms.* San Mateo, Calif.: Morgan Kaufmann, p. 80.

Grefenstette, J. J. (1986). Optimization of control parameters for genetic algorithms. *IEEE Transactions on Systems, Man and Cybernetics,* 16(1), 122.

Grefenstette, J. J. (1987). Incorporating problem specific knowledge into genetic algorithms. In L. Davis (ed.), *Genetic Algorithms and Simulated Annealing.* Los Altos, Calif.: Morgan Kaufmann, chap. 4, p. 42.

Hendrickson, J. B. (1961). Molecular geometry. I. Machine computation of the common rings. *Journal of the American Chemical Society* 83, 4537.

Holland, J. H. (1975). *Adaptation in Natural and Artificial Systems.* Ann Arbor: University of Michigan Press.

Lipton, M. and W. C. Still (1988). The multiple minimum problem in molecular modeling. Tree searching internal coordinate conformational space. *Journal of Computational Chemistry.*

Lucasius, C. B. and G. Kateman (1989). Application of genetic algorithms in chemometrics. In J. D. Schaffer (ed.), *Third International Conference on Genetic Algorithms.* San Mateo, Calif.: Morgan Kaufmann, p. 170.

Lucasius, C. B., L. M. C. Buydens, and G. Kateman (1990). Genetic algorithms for optimization problems in chemometrics. *Trends in Analytical Chemistry.* Submitted for publication.

Macura, S. and R. R. Ernst (1980). Elucidation of cross relaxation in liquids by two-dimensional NMR spectroscopy. *Molecular Physics* 41:1, 95.

Metzler, W. J., D. R. Hare, and A. Pardi (1989). Limited sampling of conformational space by the distance geometry algorithm: implications for structures generated from NMR data. *Biochemistry* 28, 7045.

Pleij, C. W. A. R. Rietveld, and L. Bosch (1985). A new principle of RNA folding based on pseudoknotting. *Nucleic Acids Research* 13, 1717.

Qian, N. and T. J. Sejnowski (1988). Predicting the secondary structure of globular proteins using neural network models. *Journal of Molecular Biology*

202:4, p. 865.

Saenger, W. (1984). *Principles of Nucleic Acid Structure.* New York: Springer-Verlag.

Smits, J. R. M., L. W. Breedveld, M. W. J. Derksen, and G. Kateman (1990). Applications of neural networks in chemometrics. In *International Neural Network Conference,* Paris.

van de Ven, F. J. M. and C. W. Hilbers (1988). Nucleic acids and nuclear magnetic resonance. *European Journal of Biochemistry* 178:1 (review).

Watson, J. D. and F. H. C. Crick (1953). A structure for deoxyribose nucleic acid. *Nature* 171, 737.

Wilson, S. R., W. Cui, J. W. Moskowitz, and K. E.D Schmidt (1988). Conformational analysis of flexible molecules: location of the global minimum energy conformation by the simulated annealing method. *Tetrahedron Letters* 29:35, p. 4373.

Wu, X. L. and R. Freeman (1989). Darwin's ideas applied to magnetic resonance. The marriage broker. *Journal of Magnetic Resonance* 85, 414.

19

Automated Parameter Tuning for Interpretation of Synthetic Images

David J. Montana

We have been developing an expert system that will automatically detect, characterize, associate and classify passive sonar signals. This system consists of a few different modules, each of which contains a large number of parameters. The importance of selecting these parameters optimally has led us to develop methods of automatically tuning these parameters which are much better than manual techniques. These automated methods have succeeded largely because they allow the system to learn (i.e., improve with experience) and adapt (i.e., change its behavior in response to changing conditions). In this chapter we describe three different automated tuning methods for three different modules of the expert system. The first involves selecting the thresholds of detection rules to yield optimal performance; the second is optimization of the parameters in an algorithm for detection and tracking of multiple signals in an image; the third is optimal selection of weights in a neural network used for classification. In all three approaches, genetic algorithms play a central role.

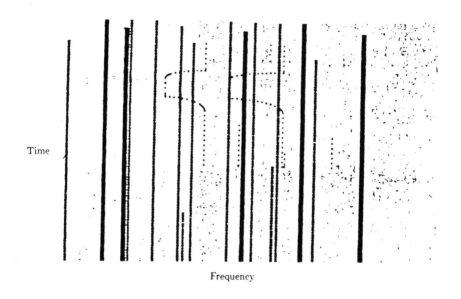

Figure 19.1: A simulated sonogram (with noise suppressed). Reprinted from Nii et al (1982).

INTRODUCTION

We start with a very brief overview of the expert system and the task it performs. (For a more detailed description of the passive sonar understanding problem and an expert system architecture suited to this problem, see Nii et al (1982).) We then examine the need for automated tuning techniques to allow the system to learn and adapt. Finally, we discuss the importance of genetic algorithms for implementing such mechanisms.

The Expert System

Our expert system operates on processed passive sonar data that it receives from an independent signal-processing module. This module transforms the incoming hydrophone data into images called sonograms. Figure 19.1

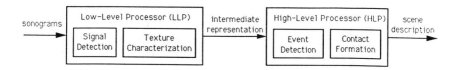

Figure 19.2: Functionality of the expert system

shows a simulated sonogram. A sensor system (such as the one from which our system receives data) produces a set of sonograms that correspond to sensing from different locations and/or in different directions. Sonar analysts read these sonogram images to identify the acoustic signatures of vessels that may be emitting sound in the area. Our expert system attempts to perform the same function.

The expert system performs four basic subtasks, which in order of increasing abstraction are (1) signal detection: determine which parts of the image correspond to emitted sound, and not just noise (known as figure-ground separation in general image analysis); (2) texture characterization: compute parameters that characterize the visual texture of the detected signals; (3) event detection: determine the location of "interesting" features of the signals; and (4) contact formation: group the signals that came from the same source into clusters (known as contacts) and attempt to classify and geographically track the contacts. These subtasks are distributed among two loosely coupled subsystems called the Low-Level Processor (LLP) and High-Level Processor (HLP). The LLP performs signal detection and texture characterization. Its primary inputs are processed data, and its primary outputs are data structures that contain signal locations and computed texture parameters. The HLP performs event detection and contact formation. It takes as inputs the LLP outputs and creates a scene description of the acoustic sources. Hence, the LLP outputs serve as an intermediate-level representation of the information in a sonogram from which the HLP forms a high-level representation. (This intermediate-level representation for sonar image understanding is analogous to the 2 1/2 D sketch for natural image understanding, see Marr 1982.) This functionality is illustrated in Figure 19.2.

The HLP is similar in architecture and functionality to HASP/SIAP (Nii et al 1982), one of the early examples of a blackboard system. Architecturally, the HLP has three main components: a global database for storing information received from the LLP plus its own inferences; rules for forming new inferences based on the data in the database; and a control structure for invoking rules at the appropriate time. We discuss particular components of the system in greater detail as required in subsequent sections.

The Need for Learning and Adaptation

Passive sonar data are very complex. Mathematical models generally do not capture all the characteristics of this data, and those that come close do not yield easily to mathematical analysis. Therefore, when building a system for analyzing sonar data, we have two distinct but interrelated tasks: to build the system and to tune it. The tuning process has received little attention in the past, despite its importance to the success of the system.

Upon examining the tuning process, we have reached three basic conclusions. First, the nature of the data requires that tuning be an ongoing process. Because of the wide range of conditions, signal types, and scenarios, any system tuned on a finite amount of data will eventually encounter a new situation for which its performance is substandard. If the system cannot improve based on this experience, then it will repeat the same mistakes in the future. As an example, consider a system tuned under low-traffic conditions. When it first encounters high traffic, it will inevitably fail. The system must subsequently learn to handle high traffic or be considered inadequate. Figure 19.3 illustrates this approach to system development and the similarities between this approach and the way human sonar analysts learn to perform the same task.

The second conclusion is that two types of learning are involved. The first type is algorithmic tuning. Playing real data through the system will highlight shortcomings and conceptual bugs in the underlying algorithms that must be fixed. At the present time, this type of learning is best done by the human developers with the aid of tools on the machine. The second type of learning is parameter tuning. Our system contains a large number of parameters whose settings greatly influence performance. Choosing the best values for these parameters is generally difficult for a human for a number of reasons, including (1) interactions between parameters, (2) the difficulty humans have mentally juggling large amounts of data to make statistically based decisions, (3) the long time required to evaluate a single set of parameters, and (4) the difficulty of changing the parameters to meet new system specifications. We have therefore been developing methods

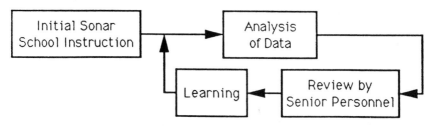

(a) Process of Professional Development for Human Analyst

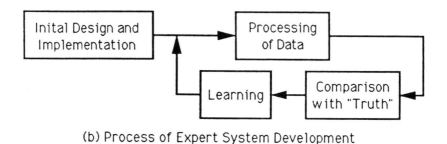

(b) Process of Expert System Development

Figure 19.3: Comparison of development paths of a human sonar analyst and our expert system.

whereby the machine can learn appropriate parameter settings. We discuss these in the next three major sections of this chapter.

The third conclusion is that parameter adaptation (i.e., changing parameter values in response to changing conditions) can greatly improve system performance. For example, the parameter settings that are optimal for high-traffic data are different from those that are optimal for low-traffic data. Hence, a system that uses one appropriate set of parameter values for high traffic and another for low traffic will outperform a system that uses a single set of parameter values under both conditions.

The Importance of Genetic Algorithms

Although (as we will see below) we have used different approaches to automating the tuning of different pieces of our expert system, in each case we have used a genetic algorithm for performing optimization. The properties of genetic algorithms that have made them well suited to our applications are the following. First, they generally find nearly global optima in complex spaces. This is important because the search spaces for our problems are highly multimodal, a property that leads hill-climbing algorithms to get stuck in local optima. Second, genetic algorithms do not require any form of smoothness. This is important because in two cases (rule threshold optimization and tracking parameter optimization) the search space is discontinuous. Third, considering their ability to find global optima, genetic algorithms are relatively fast, especially when tuned to the domain on which they are operating. (They are especially fast when distributed across different machines on a network as discussed later in this chapter .) In all cases, the speed of the optimization algorithm is one of the prime factors determining the practicality of the approach. Fourth, tuning a genetic algorithm for a particular domain is relatively easy because genetic algorithms consist of a general format with some variable components chosen to meet the needs of the domain. The five variable components are

1. A way of encoding solutions to the problem on chromosomes.

2. An evaluation function that returns a rating for each chromosome given to it.

3. A way of initializing the population of chromosomes.

4. Operators (e.g., mutation and crossover) that may be applied to parents when they reproduce to alter their genetic composition.

5. Parameter settings for the algorithm, the operators, etc.

If: (< (average-kludginess signal) 20000)
      (> (lossage-derivative signal time) 0.01)
      (> (friendliness-at-time signal time) 5.5)
Then: (declare-event-type-foo signal time)

Figure 19.4: Example event detection rule packet.

OPTIMIZATION OF THRESHOLDS IN EVENT DETECTION RULES

Some particularly important rules in the HLP are those that detect events. Their function is to decide whether or not a certain signal has a certain type of event at a certain time. They thus perform pattern recognition, distinguishing between positive and negative examples of different types of events. An example of an event detection rule is shown in Figure 19.4. Note that the conditions of the rule consist of tests of real-valued functions of the database against real-valued thresholds. These thresholds are parameters whose values can be varied to optimize detection performance. In this section we examine how to automate the process of selecting values for these thresholds.

We start by discussing ROC curves. We then describe the two key components of our optimization procedure: (1) genetic algorithms and (2) windowing. Finally, we examine the results and benefits of using this automated approach.

Receiver Operating Characteristic (ROC) Curves

For the detection problem, there are two basic types of errors: a missed detection (classifying a positive example as a negative one), and a false alarm (classifying a negative example as a positive one). These two types of errors give rise to two different and competing measures of detection performance: probability of detection (P_d), the fraction of positive examples classified correctly, and probability of false alarm (P_f), the fraction of negative examples classified incorrectly. Many detection algorithms have parameters that can be changed to yield different performance and thus a different pair of performance measures (P_d, P_f). A realizable (P_d, P_f) is called Pareto optimal if there exists no other realizable (P_{d1}, P_{f1}) for which $P_{d1} \geq P_d$ and $P_{f1} \leq P_f$ and at least one of these inequalities is a strict

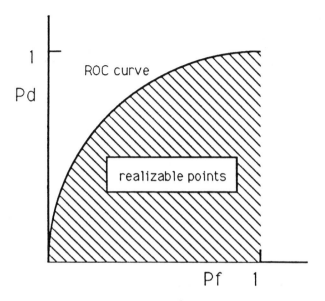

Figure 19.5: Example of an ROC curve.

inequality. The set of all Pareto optimal pairs (P_d, P_f) forms a curve in $P_d - P_f$ space called an ROC curve. An example of an ROC curve is shown in Figure 19.5.

Genetic Optimization of Rule Thresholds

In this section we describe the various components of our genetic algorithm for rule threshold optimization. The algorithm is implemented using an early version of Lawrence Davis's OOGA code.

Encoding: We use a real-valued encoding scheme instead of the traditional binary one. An individual consists of a list of the thresholds in the order in which they appear in the rule. The possible values that a threshold may take are both range-limited and quantized. As one of the functions of our rule editor, the developer specifies the maximum value, minimum value, and step size for a particular threshold. The developer picks these parameters based on knowledge of typical values of the corresponding statistic. This approach is necessary because the statistics used in the rules have a wide range of typical values. For instance, one statistic may generally have

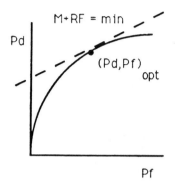
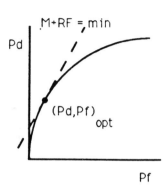

Figure 19.6: Two different values of R yielding two different optimal points along the ROC curve.

values between 5 and 15 while another may typically have values between 0.001 and 0.004. Knowing this information makes the genetic algorithm much more effective because it does not have to waste its time searching out of a statistic's range or on a scale that is insignificant with respect to the statistic. Shaefer (1987) describes a genetic optimization technique called ARGOT which automatically handles the large dynamic range of parameters by adapting the representation during a run.

Evaluation Function: A training database of positive and negative examples exists for each event type. (How we construct this database is the topic of the next subsection, "Windowing.") An automatic scoring function loops through all examples of the appropriate event type in the training database and counts the number of missed detections, M, and the number of false alarms, F, for a given set of rule thresholds. Let N_p be the total number of positive examples and N_n be the total number of negative examples. Then an estimate of P_d is $1 - M/N_p$, and an estimate of P_f is F/N_n. Note that the training database is stacked with particularly difficult cases owing to the windowing procedure (see "Windowing"). Hence, the calculated P_d and P_f are not good absolute estimates of general rule performance. In particular, the calculated P_f is orders of magnitude greater than the actual P_f of the rules. However, these scores do provide a good way to compare relative performance of rules, and they can be computed fairly quickly and effortlessly.

The evaluation function is defined as $M + RF$, where R is a parameter

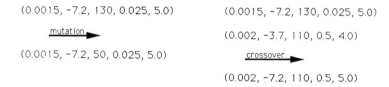

Figure 19.7: The genetic operators mutation and crossover.

whose value is selected by the developer. The choice of R specifies the operating point on the ROC curve (assuming that the genetic algorithm does indeed find global optima). Figure 19.6 shows how two different choices of R lead to two different points on the ROC curve. Allowing the developer to choose the value of R provides him with the capability of choosing an arbitrary operating point on the ROC curve. This capability is crucial to the success of the system. (Schaffer (1985) describes a system called VEGA which tries to find multiple points distributed over the ROC curve rather than just a single point.)

Initialization: The threshold settings for each individual in the initial population are randomly selected from the admissible set.

Operators: We use the two basic genetic operators, crossover and mutation, suitably modified for the particular representation scheme. Our mutation operator creates a child that is the same as the parent in all locations except one or more randomly selected ones. The threshold values of the child in these locations are chosen randomly from their allowable sets (see Figure 19.7a). (The property that the child must be different from the parent is what Davis (1989) calls "guaranteed".)

Our crossover operator takes two parents and creates a single child. It selects each threshold value of the child by randomly choosing one of the two parents and using the corresponding threshold value in that parent (see Figure 19.7b). This is an example of uniform crossover; we have chosen to use this type of crossover based on results reported in Syswerda (1989).

Parameters: The values of a number of parameters can greatly influence the performance of the algorithm. Some of the important parameters are as follows.

PARENT-SCALAR: This parameter determines with what probability each individual is chosen as a parent. The second-best individual is

PARENT-SCALAR times as likely as the best to be chosen, the third-best is PARENT-SCALAR times as likely as the second-best, and so forth. The value was linearly interpolated between 0.92 and 0.89 over the course of a run.

OPERATOR-PROBABILITIES: This list of parameters determines with what probability each operator in the operator pool is selected. These values were initialized so that mutation and crossover had equal probabilities of selection. An adaptation mechanism changes these probabilities over the course of a run to reflect the performance of the operators in a manner described in Davis (1989).

POPULATION-SIZE: This self-explanatory parameter was set to 50.

GENERATION-SIZE: This parameter tells how many children to generate for each iteration (and how many current population members to delete). It was set to one. In the terminology of Syswerda (1989), this makes our genetic algorithm a steady-state genetic algorithm rather than a generational replacement algorithm. The advantage of having the generation size be as small as possible is to be able to immediately incorporate better individuals created via reproduction into the reproductive process as potential parents.

Windowing

It may be hard to find many positive examples of a particular event type, but there is no shortage of negative examples. In fact, so many negative examples are encountered in even a relatively short time slice of continuous sonar data that it is computationally infeasible to optimize the rule thresholds with all these examples in the training database. We therefore require a way to select a training database which contains only a small subset of all the examples encountered but which is representative of the full set of examples. Windowing (Quinlan 1979) is a process which allows us to choose such a subset.

Windowing works according to the following steps: (1) Randomly select a small subset of the examples called the window (which we call the training database); (2) Train the algorithm on the window; (3) Search through the full training set for incorrectly classified examples and add them to the window if they are not already there; (4) If new examples were added to the window, repeat from step (2). This process is illustrated in Figure 19.8. (Note the similarity in form between the windowing procedure shown in Figure 19.8 and the general system development process shown in Figure 19.3. Windowing is an example of this general approach.)

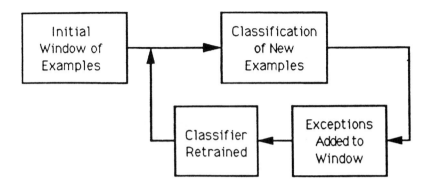

Figure 19.8: Windowing procedure.

We have devised a user interface (described in Montana 1990) that allows a sonar analyst to view the decisions of the rules and add misclassifications into the training database. We have thus been able to perform numerous passes through the windowing cycle. As we do so, the training database becomes more representative of the full range of data, and hence event detection performance continually improves: that is the system learns with experience. Results concerning this learning process are given in the following section.

Results and Benefits

We have discovered a number of results about the rule threshold optimization procedure described above. These are described in detail in Montana (1990) and are summarized here. One result involves how the genetic algorithm scales as a function of the number of thresholds in a rule; we have found that this function is approximately linear and that the number of evaluations required is approximately 100 to 150 times the number of thresholds. Our quantization of the thresholds is such that each threshold on the average represents around five binary dimensions. Hence, the scaling factor is 20 to 30 times the binary dimension.

A second result concerns a comparison between automated and manual tuning. An early version of the system has rule thresholds set by hand by human developers. A later version of the system has rule thresholds determined using the automated technique described in this section. A comparison of the results is shown in Figure 19.9. Note that (1) the automated

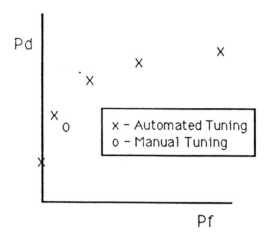

Figure 19.9: Automated tuning versus manual tuning.

approach generated an operating point that is better in a Pareto sense than the manual technique, and (2) the automated approach generated a full ROC curve of operating points simply by choosing different tradeoffs between P_d and P_f and running the genetic optimizer for each tradeoff (note that the time-consuming part of the automated approach is generating the training database via windowing), while the manual approach generated only a single point (and would require starting from scratch to generate a different point).

By enabling the user to easily change the operating point on an ROC curve, the automated technique provides a number of benefits. One is the ability to adapt to changes in system specifications. A second benefit is the ability to have event detection rules with different confidences (derived by changing the tradeoff between P_d and P_f) in the same system (see Montana 1990). In this case, higher level modules can make the ultimate detection decisions based on the confidence associated with detections as well as the existence or lack of confirming evidence from other sources. A third benefit is the ability to adapt to changing conditions. For instance, high-traffic conditions require a lower P_f than low-traffic conditions; hence, tuning rule thresholds for high traffic means putting greater weight on P_f than for low traffic. The system can switch in the appropriate rule thresholds as required.

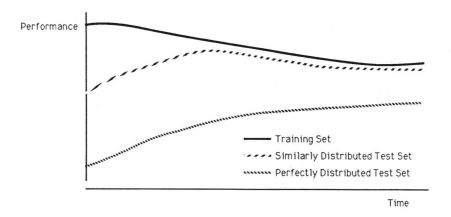

Figure 19.10: Evolution of performance with time.

A final result concerns how the system improves with experience, that is learns. Figure 19.10 presents a graph of system performance versus time on three different types of data. The training data are a randomly selected subset of the full training database. The similarly distributed test data are the data from the full training database that were not chosen as part of the training data. The perfectly distributed training data in theory are a set of examples representative of all possible signals and conditions, but in practice they represent a large amount of independent data. True performance is measured with respect to the perfectly distributed training data. Learning occurs as the training data become more and more like perfectly distributed data, and hence true performance improves.

OPTIMIZATION OF SIGNAL DETECTION PARAMETERS

The signal detection module detects the presence of (possibly simultaneous) signals in a sonogram and determines their location in the image. Most of the details of the algorithm do not matter for the purposes of this chapter.

Even so, we do need to convey some understanding of the large number of parameters in the algorithm, the variety of functions of these parameters, and how these parameters interact. There are four basic parts to the algorithm, each with its own associated parameters: (1) track creation: decide where to start new tracks; (2) track extension: decide how existing tracks proceed through the sonogram (this includes a track smoothing filter with situation-dependent filter parameters); (3) track validation: decide which sections of which tracks to consider to be a true signal rather than noise; and (4) track deletion: decide where the tracks end. Selecting these parameters appropriately is crucial to the success of the algorithm. We have been able to consolidate into 20 significant parameters what once was a larger number of partially redundant parameters. In this section we discuss a method for automatically finding an optimal set of parameters.

The genetic algorithm used to optimize these parameters is the same as that used for rule threshold optimization with a few key differences, which we discuss below. The first is the scoring function used to evaluate a set of parameters, which is similar but more complex than the evaluation criterion for rule thresholds. The second is the fact that the evaluation computations are distributed across multiple computers on a network. We conclude the chapter with a discussion of results and benefits.

The Scoring Function

As is the case for rule threshold optimization, we need a training database against which to score the performance of a set of signal detection parameters. We call this the "truth tracks" database, a phrase that refers to the fact that we are not just detecting the presence of signals but are also tracking their location in the sonogram. Because of the need to track signals, examples in the truth tracks database must contain desired outputs that are much more than simply yes or no (i.e., detect or reject). Instead, they consist of (1) a list of all signals, (2) for each signal, a list of all pixels in which it is visible, (3) a characterization of each signal, and (4) locations of events which should be detected by the event detection rules. (Signal characterizations and events are needed to help weight different signal detection errors according to their importance.) To get all this information, especially that at the pixel level, is very difficult and tedious. To make the process of "truthing" the data much simpler, we have developed a graphical, menu-driven tool for entering and editing data in the truth tracks database.

The scoring function compares the outputs of the signal detection module on some gram regions with the truth tracks for these gram regions to compute a single performance score. This score is analogous to the $M + RF$ score for event detection rules. However, the signal detection score is much more complex for the following three reasons. First, in addition to the basic tradeoff between detections and false alarms, we must consider other performance criteria, such as how close the computed track locations are to the truth track locations. Second, there are great differences in the importance of different signals and pieces of signals. For example, it is particularly important to detect and track correctly the pieces of signals near events. This means the scoring function must weight all errors based on their importance. Third, the outputs of the signal detection module are not an end in themselves but rather are data to feed into higher-level processing modules. This makes scoring signal detection performance partially subjective in the sense that it depends on opinions of how different types of errors affect the higher-level processing.

The score is computed in two stages. The first stage is to compute a correspondence map between pixels of truth tracks and pixels of computed tracks. This map is defined as follows. A pixel of a truth track corresponds to a pixel of a computed track if and only if (1) the computed track pixel is the closest such pixel to the truth track pixel, (2) the truth track pixel is the closest such pixel to the computed track pixel, and (3) the computed track is tracking the truth track at some point. This leaves some truth track pixels unassociated with computed track pixels and vice versa. We call a truth track or computed track unassociated if all its pixels are unassociated.

The second stage of the evaluation computation is to calculate from the correspondence map a small number of performance measures (analogous to P_d and P_f for event detection rules) which can then be combined using a weighted sum to give a single score. The developer chooses the weighting coefficients before the start of an optimization run. We currently use the following six performance measures.

- Missed pixels: weighted sum (with a weighting scheme based on importance) over all truth tracks of unassociated pixels.

- Added pixels: weighted sum over associated computed tracks of unassociated pixels.

- Trash pixels: sum over unassociated computed tracks of all pixels.

- Breaks: weighted sum over associated truth tracks of the number of corresponding tracks minus one.

- Jumps: weighted sum over associated computed tracks of the number of corresponding tracks minus one.

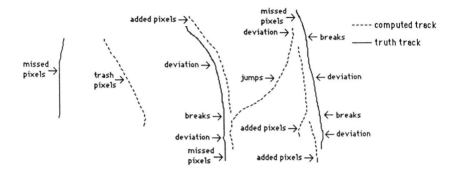

Figure 19.11: Examples of contributions to the different performance measures.

- Deviation: a weighted sum over associated truth track pixels of the square of the distance from its corresponding computed track pixel.

Figure 19.11 presents examples of contributions to each of these scores.

Distributed Evaluations

There are two reasons for distributing the evaluations across machines. First, the signal detection code runs only on SUN computers, whereas the scoring and genetic algorithm code run only on Symbolics computers. The easiest solution to this incompatibility is to have the evaluation function on a Symbolics tell a SUN to run the signal detection algorithm with parameters as given in a given individual. The evaluation function then reads the computed tracks to the Symbolics for the scoring function. A second, more basic reason for distributing the evaluations is speed. Running the signal tracking algorithm on a gram piece typically takes a few minutes, and the scoring function can take an additional minute. If, as an example, we are using three gram pieces, then evaluating each individual might take 10 to 15 minutes, and evaluating many individuals quickly adds up to unmanageably long runs.

We distribute the processing as follows. There are two resource pools, one for SUN and one for Symbolics. Each resource pool contains a list of available hosts and a queue of processes waiting for a host. As hosts become available, processes are taken off the queue and allowed to run. The evaluation function generates one process for each individual and gram piece. Each process first seizes a SUN from the resource pool to run the signal detection algorithm on its gram piece with parameters given by its individual. After this finishes, the process seizes a Symbolics to score the outputs of the signal detection module. When the scoring is finished, the process increments the cumulative score for its individual. This procedure is illustrated in Figure 19.12.

One interesting issue is how the degree of parallelism affects choices of parameters for the genetic algorithm. The only parameter we have found necessary to change with the degree of parallelism is the generation size. For the reason explained in "Genetic Optimization of Rule Thresholds," making the value of this parameter smaller means that the genetic algorithm requires fewer evaluations before convergence. However, with parallelism available, making the generation size smaller also means increasing the average time needed per evaluation. The reason for this is the need for synchronization between generations. Before the genetic algorithm can create the individuals of a new generation, it needs to finish evaluating those of the old generation. During this synchronization period, the evaluation function cannot achieve full parallelism. Since the time needed for synchronization does not change with the generation size, the larger the generation size the larger the average amount of parallelism. When running with two grams in the training database, three SUN's, and one Symbolics, we have found five to be a good value for the generation size.

(Note that the optimal solution (which we have yet to implement) to the tradeoff between using good individuals in the reproduction process as soon as possible and obtaining maximal parallelism is to do away with the concept of generations. Instead, let the reproduction and evaluation processes be asynchronous, that is (1) create a new individual for evaluation as soon as there is a free resource to start the evaluation process and (2) as soon as an individual is finished being evaluated, insert it into the population, eliminating the worst population member to make room. In this case, computing resources are always fully utilized while individuals can reproduce as soon as they are evaluated.)

Results and Benefits

The signal detection parameter optimization procedure is a more recent development than the rule threshold optimization procedure. Hence, there

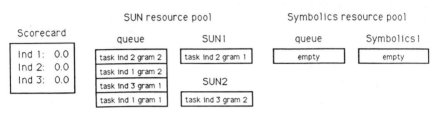

(a) Configuration immediately after starting evaluations

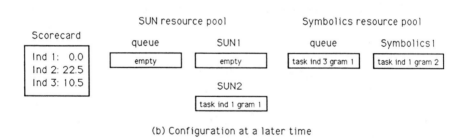

(b) Configuration at a later time

Figure 19.12: Distributing the genetic algorithm.

are few results documenting its performance. We have experimented with it on a relatively small training database with the following results. First, it takes on the order of 500 evaluations for the genetic algorithm to converge. (This is a rule of thumb and has not been statistically validated.) This corresponds to 25 evaluations per real-valued dimension or five evaluations per binary dimension.

Second, with appropriate weights, the optimization routine produces parameters that are better than the manually tuned parameters on the training database in the following respects: (1) its overall score is better (as it should be or else the optimization is not finding a global optimum), (2) each of its six individual scores is better, and (3) most importantly, developers familiar with the expert system agree that its outputs are better. (Recall that there is no truly objective measure of performance because these are only intermediate results.) Work continues to determine whether we can get similar improvements with a larger training database.

Another benefit is the ability to easily alter the tradeoff between the different performance measures in computing a single score as well as the relative weightings of the different signals and pieces of signals in computing the performance measures. Plans for this capability include different parameter sets for different conditions and multiple signal detection routines running on the same data with different parameter sets (thus tuned for detecting particular types of signals).

TRAINING OF NEURAL NETWORKS FOR TEXTURE CHARACTERIZATION

The present texture characterization module computes known and understood parameters of a signal to provide clues to the higher level modules. We have been experimenting to see if we can instead perform classification directly from the texture. We have used neural networks as an underlying classification structure and have trained them on a database of examples of different types of signal textures. Training a neural network means selecting its free parameters so as to optimize its performance (measured by a chosen evaluation function) on the training database. The standard method for training neural networks is called backpropagation (Rumelhart 1986) and is a variant on gradient search. However, for our problem the complexity of the search space caused backpropagation to continually get stuck in local optima far in value from the global optimum. For obtaining better optima, we have used a genetic algorithm for training neural networks. In this section we describe this genetic training algorithm and the results obtained with it. The material in this section also appears in Montana and Davis (1989).

Neural Networks

Neural networks generally consist of five components:

1. A directed graph known as the network topology whose arcs we call links.

2. A state variable associated with each node.

3. A real-valued weight associated with each link.

4. A real-valued bias associated with each node.

5. A transfer function for each node which determines the state of a node as a function of (a) its bias ψ, (b) the weights, w_i of its incoming links, and (c) the states, x_i, of the nodes connected to it by these links. This transfer function usually takes the form $f(\sum w_i x_i - \psi)$ where f is either a sigmoid or a step function.

A feedforward network is one whose topology has no closed paths. Its input nodes have no arcs to them, and its output nodes have no arcs away from them. All other nodes are hidden nodes. When the states of all the input nodes are set, all the other nodes in the network can set their states as values propagate through the network. The operation of a feedforward network consists of calculating outputs given a set of inputs in this manner. In a layered feedforward network, any path from an input node to an output node traverses the same number of arcs. The nth layer of such a network consists of all nodes that are n arc traversals from an input node. A hidden layer is one which contains hidden nodes. Such a network is fully connected if each node in layer i is connected to all nodes in layer $i + 1$ for all i. We have restricted ourselves to working with layered, feedforward networks. Figure 19.13 shows such a network.

The Genetic Algorithm

We now describe the five components of our genetic algorithm for training neural networks.

Encoding: The weights (and biases) in the neural network are encoded in order as a list (see Figure 19.13).

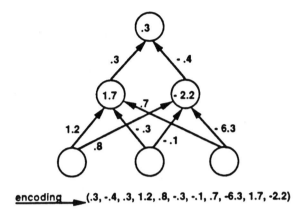

encoding ──▶ (.3, -.4, .3, 1.2, .8, -.3, -.1, .7, -6.3, 1.7, -2.2)

Figure 19.13: Encoding a network on a Chromosome.

Evaluation Function: The weights on the chromosome are assigned to the links in a network of a given architecture, the network is run over the training set of examples, and the sum of the squares of the errors is returned from each example.

Initialization: The entries (i.e., weights) of the initial members of the population are chosen at random with a probability distribution given by $e^{-\|x\|}$, that is a two-sided exponential distribution with a mean of 0.0 and a mean absolute value of 1.0. This is different from the initial probability distribution of the weights generally used in backpropagation, which is a uniform distribution between -1.0 and 1.0. Our probability distribution reflects the empirical observation that optimal solutions tend to contain predominantly weights with small absolute values but that they can have weights with arbitrarily large absolute values. We therefore seed the initial population with genetic material that allows the genetic algorithm to explore the range of all possible solutions but that tends to favor the most likely solutions.

Operators: We created a large number of different types of genetic operators. The goal of most of the experiments we performed was to find out how different operators perform in different situations and thus to be able to select a good set of operators for the final algorithm. The operators can be grouped into three basic categories: mutations, crossovers, and gradients. A mutation operator takes one parent and randomly changes some of the entries in its chromosome to create a child. A crossover operator takes two parents and creates one or two children containing some of the genetic material of each parent. A gradient operator takes one parent and produces a child by adding to its entries a multiple of the gradient with respect to the evaluation function. We now discuss each of the operators individually, one category at a time.

UNBIASED-MUTATE-WEIGHTS: For each entry in the chromosome, this operator will with fixed probability $p = 0.1$ replace it with a random value chosen from the initialization probability distribution.

BIASED-MUTATE-WEIGHTS: For each entry in the chromosome, this operator will with fixed probability $p = 0.1$ add to it a random value chosen from the initialization probability distribution. We expect biased mutation to be better than unbiased mutation for the following reason. Right from the start of a run, the parents chosen tend to be better than average. Therefore, the weight settings in these parents tend to be better than random settings. Biasing the probability distribution by the present value of the weight should then give better results than a probability distribution centered on zero.

MUTATE-NODES: This operator selects N noninput nodes of the network that the parent chromosome represents. For each ingoing link to these N nodes, the operator adds to the link's weight a random value from the initialization probability distribution. It then encodes this new network on the child's chromosome. The intuition here is that the ingoing links to a node form a logical subgroup of all the links in terms of the network's operation. By confining its changes to a small number of these subgroups, it will make its improvements more likely to result in a good evaluation. In our experiments, $N = 2$.

MUTATE-WEAKEST-NODES: This operator uses a quantity we called node strength. Node strength is different from the concept of error used in backpropagation. For example, a node can have zero error if all its output links are set to zero, but such a node is not contributing anything positive to the network and is thus not a strong node. We define the strength of a hidden node in a feedforward network as the difference between the evaluation of the network intact and the evaluation of the network with that node lobotomized (i.e., with its output links set to zero). We have devised a more efficient way to calculate node strength, but it is not discussed here.

The operator MUTATE-WEAKEST-NODES takes the network that the parent chromosome represents and calculates the strength of each hidden node. It then selects the M weakest nodes and performs a mutation on each of their ingoing and outgoing links. This mutation is unbiased if the node strength is negative and biased if the node strength is positive. It then encodes this new network on a chromosome as the child. The intuition behind this operator is that some nodes are not only not very useful to the network but may actually be hurting it. Performing mutation on these nodes is more likely to yield bigger gains than mutation on a node that is already doing a good job. Note that since this operator will not be able to improve nodes that are already doing well it should not be the only source of diversity in the population. In our experiments, $M = 1$.

CROSSOVER-WEIGHTS: This operator puts a value into each position of the child's chromosome by randomly selecting one of the two parents and using the value in the same position on that parent's chromosome.

CROSSOVER-NODES: For each node in the network encoded by the child's chromosome, this operator selects one of the two parents' networks and finds the corresponding node in this network. It then puts the weight of each ingoing link to the parent's node into the corresponding link of the child's network. The intuition here is that networks succeed because of the synergism between their various weights, and this synergism is greatest among weights from ingoing links to the same node. Therefore, as genetic material gets passed around, these logical subgroups should stay together.

CROSSOVER-FEATURES: Different nodes in a neural network perform different roles. For a fully connected, layered network, the role that a given node can play depends only on which layer it is in and not on its position in that layer. In fact, we can exchange the role of two nodes A and B in the same layer of a network as follows. We can loop over all nodes C connected (by either an ingoing or outgoing link) to A (and thus also to B) and we can exchange the weight on the link between C and A with that on the link between C and B. Ignoring the internal structure, we see that the new network is identical to the old network. That is, given the same inputs they will produce the same outputs.

The child produced by the previously discussed crossovers is greatly affected by the parents' internal structures. The CROSSOVER- FEATURES operator reduces this dependence on internal structure by doing the following. For each node in the first parent's network, it tries to find a node in the second parent's network which is playing the same role by showing a number of inputs to both networks and comparing the responses of different nodes. It then rearranges the second parent's network so that nodes playing the same role are in the same position. At this point, it forms a child in the same way as CROSSOVER-NODES. The greatest improvement gained from this operator over the other crossover operators should come at the beginning of a run before all the members of a population start looking alike.

HILLCLIMB: This operator calculates the gradient for each member of the training set and sums them together to get a total gradient. It then normalizes this gradient by dividing by the magnitude. The child is obtained

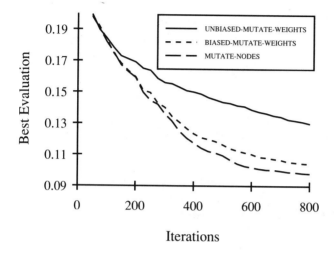

Figure 19.14: Comparison of mutations.

from the parent by taking a step in the direction determined by the normalized gradient of size step-size, where step-size is a parameter that adapts throughout the run in the following way. If the evaluation of the child is worse than the parent's, step-size is multiplied by the parameter step-size-decay = 0.4; if the child is better than the parent, step-size is multiplied by step-size-expand = 1.4. This operator differs from backpropagation in the following ways: (1) weights are adjusted only after calculating the gradient for all members of the training set, and (2) the gradient is normalized so that the step size is not proportional to the size of the gradient.

Parameters: The parameters were the same as for rule threshold optimization.

Results

We have run a series of experiments discussed in detail in Montana and Davis (1989). We summarize the results here. First, comparing the three general-purpose mutations resulted in a clear ordering (see Figure 19.14):

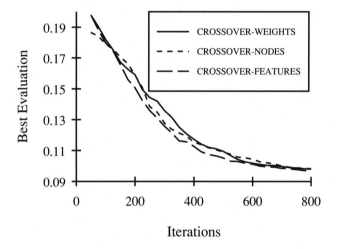

Figure 19.15: Comparison of crossovers.

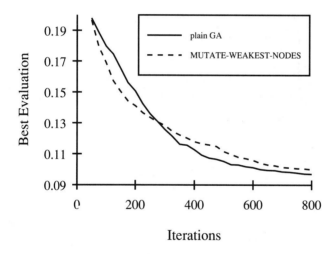

Figure 19.16: Effect of MUTATE-WEAKEST-NODES.

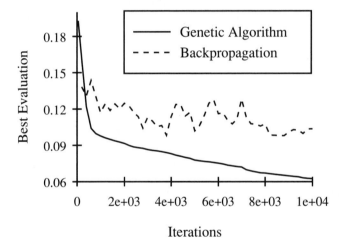

Figure 19.17: Comparison of genetic algorithm with backpropagation.

1. MUTATE-NODES,

2. BIASED-MUTATE-WEIGHTS, and

3. UNBIASED-MUTATE-WEIGHTS.

Second, a comparison of the three crossovers produced no clear winner (see Figure 19.15). Third, when MUTATE-WEAKEST-NODE was added to a mutation and crossover operator, it improved performance only at the beginning; after a certain small amount of time, it decreased performance (see Figure 19.16). Fourth, a hill-climbing mode with just a HILLCLIMB operator and a population of size two gave temporarily better performance than MUTATE-NODES and CROSSOVER-NODES but would quickly get stuck at a local minimum. Finally, a genetic training algorithm with operators MUTATE-NODES and CROSSOVER-NODES outperformed back-propagation on our problem (see Figure 19.17).

CONCLUSIONS

Automated parameter tuning has provided a big boost to the performance of our expert system and promises to do so even more in the future. The expert system consists of four different modules, and we have applied automated parameter tuning to three of them: (1) optimizing rule thresholds for the event detection module, (2) optimizing algorithm parameters for the signal detection module, and (3) optimizing the free parameters of a neural network for the texture characterization module. In each case, genetic algorithms have played a key role.

ACKNOWLEDGMENTS

Thanks are due to the following people: David Davis, for his tutelage in the field of genetic algorithms, his recognition of the potential for genetic algorithms for these parameter optimization problems, and the use of his OOGA code; Steve Milligan, for suggesting the ideas of optimizing rule thresholds and applying neural networks to texture characterization; Fred White, for writing code to support the rule threshold optimization process; and Ken DeJong, for his useful comments.

REFERENCES

Davis, L. (1989). Adapting operator probabilities in genetic algorithms. *Proceedings of the Third International Conference on Genetic Algorithms,* pp. 61–69.

Marr, D. (1982). *Vision.* San Francisco: W.H. Freeman and Co.

Montana, D. J. (1990). Empirical learning using rule threshold optimization for detection of events in synthetic images. To appear in *Machine Learning.*

Montana, D. J. and Davis, L. (1989). Training feedforward neural networks using genetic algorithms. *Proceedings of the Eleventh International Joint Conference on Artificial Intelligence,* pp. 762–767.

Nii, H. P. et al (1982). Signal-to-symbol-transformation: HASP/SIAP case study. *AI Magazine* 3(2), pp. 23–35.

Quinlan, J. R. (1979). Discovering rules by induction from large numbers of examples: a case study. In D. Mitchie (ed.), *Expert Systems in the Microelectronic Age.* Scotland: Edinburgh University Press.

Rumelhart, D., Hinton, G. and Williams, R. (1986). Learning representations by backpropagating errors. *Nature* 323, pp. 533–536.

Schaffer, J. D. (1985). Multiple objective optimization with vector evaluated genetic algorithms. *Proceedings of the First International Conference on Genetic Algorithms,* pp. 93–100.

Shaefer, C. G. (1987). The ARGOT strategy: adaptive representation genetic optimizer technique. *Proceedings of the Second International Conference on Genetic Algorithms,* pp. 50–58.

Syswerda, G. (1989). Uniform crossover in genetic algorithms. *Proceedings of the Third International Conference on Genetic Algorithms,* J. David Schaffer, editor, pp. 2-9. Morgan Kaufmann.

20

Interdigitation: A Hybrid Technique for Engineering Design Optimization Employing Genetic Algorithms, Expert Systems, and Numerical Optimization

David J. Powell, Michael M. Skolnick, and Siu Shing Tong

INTRODUCTION

The ability of engineers to produce optimal designs has been severely limited by the techniques available for design optimization. Typically, much of the development effort has focused on simulation programs to evaluate design parameters. It is the design, implementation, and evaluation of these programs that most directly call upon the engineer's expertise. One problem currently being addressed is how to use the information provided by the simulations in the iterative process of searching the parameter space for better designs. Given an evaluation of a setting of the parameters governing a design, on what basis should the choice be made of the "best" parameters to evaluate next? Because design spaces are combinatorially explosive and the time in which to develop a new design is limited, relatively few design points can be evaluated. Furthermore, since simulation programs rather than explicit analytic functional forms are typically used to model the design, the engineer's understanding of the parameter search space is limited (e.g., as to the numbers and types of regions corresponding to local optima).

The problem is that engineering knowledge sufficient to produce sound simulation programs is difficult enough to achieve, but knowledge of the space of feasible design points is much more limited. This lack of knowledge is the reason that we need automated procedures for iterative design.

Traditionally, the techniques of numerical optimization have been used to guide the search through parameter space (Vanderplaats 1985). More recently, the technology of expert systems has been integrated into this process (Tong 1986). Both approaches to the problem have fundamental strengths but tend to suffer in constrained optimization problems, where there are large, nonlinear spaces and there is incomplete domain-dependent knowledge to guide the search. It is in these situations that the genetic algorithm is designed to work.

A unique combination of expert systems, numerical optimization, and genetic algorithms offers all the advantages of available optimization techniques while offsetting their disadvantages. This "interdigitized" hybrid technique for design automation couples all three classes of techniques tightly and switches between them within a design run. (The term *interdigitized* is derived from the image of two clasped hands with fingers intertwined.) This technique is very general and allows the engineer to concentrate on the design problem without having to worry about the selection and fine tuning of optimization algorithms.

The interdigitized technique described here has been validated against a diverse set of engineering design problems that have proven to be unsolvable using any single numerical optimization technique. It has also been used in the design of General Electric Company's DC motors and in the design of a turbine for GE's largest commercial engine—where the final design surpassed the efficiency of all turbines designed using other optimization techniques. The "Applications" section of this chapter discusses the test set, which is also presented as a proposed standard for anyone working with genetic algorithms and design automation. Also discussed in the "Applications" section is the design of the turbine for the large GE engine.

SURVEY OF OPTIMIZATION TECHNIQUES

Three optimization techniques frequently applied in design automation are expert systems, numerical optimization, and genetic algorithms. The discussion that follows summarizes the major advantages and disadvantages of each.

Expert Systems

Expert systems codify knowledge about a domain in the form of IF THEN rules that are manipulated by forward and backward inference techniques to provide solutions to design problems. Expert systems have been used to develop optimal designs in a number of domains, including cooling fans (Tong 1986), VLSI circuits (Jabri 1987), and bridges (Adeli 1986). Expert systems offer many advantages: they make use of the engineer's domain knowledge; they provide solutions to design problems efficiently; and they explain how those solutions were obtained. They also have disadvantages: they require rules that describe a domain completely; they cannot adapt readily to change within a domain (e.g., changing output constraints); and they are domain-dependent. Even for a design of minimal complexity, rule completeness is not possible because the engineer does not understand the design space completely. Acquiring the knowledge that the engineer does have is difficult because of the mismatch between the way engineers express their knowledge and the format required by rule representation in the computer. Knowledge acquisition is always hindered by the inability of humans to express all their knowledge and by the errors introduced as knowledge is transferred from a design engineer to a knowledge engineer (Quinlan 1987; Zhou 1987).

Numerical Optimization

Numerical optimization uses gradient approximations to calculate search directions leading to an optimum. For nonlinear constrained optimization, the techniques of sequential linear programming (SLP), sequential quadratic programming (SQP), the modified method of feasible directions (MMFD) (Vanderplaats 1985), and the generalized reduced gradient method (GRG) (Gabriele 1988) are the best numerical techniques. They have been successfully applied to optimization problems in a number of domains, including chemical process design, aerodynamic optimization, nonlinear control systems, mechanical component design, and structural design (Vanderplaats 1985). The advantages of numerical optimization are its mathematical underpinnings, its general applicability to engineering designs, and its wide application base. The disadvantages are its inability to exploit domain knowledge, its extreme sensitivity to both problem formulation (Gero 1988) and algorithm selection, its need for large amounts of computational effort, and its assumptions that the design variables are independent and the parameter space is continuous. Numerical optimization techniques are good at "exploitation" but not "exploration" of the parameter space. They are successful at exploitation because they focus on the immediate area around the current design point, using local gradient calculations to move to a better design. Since no attempt is made to explore all the regions of the parameter space, however, numerical optimization can easily be trapped in local optima or by constraints in a region of the parameter space far from the optimal design (Booker 1987).

Genetic Algorithms

Genetic algorithms take an initial set, or population, of design points and manipulate that set with the genetic operators of selection, crossover, and mutation to arrive at an optimal design. Seeding is the process of providing the initial set of design points. Although the seeding is typically performed by random selection, some systems such as EnGENEous have used past designs from the parameter space to provide a portion of the initial design set (Powell et al. 1989). Genetic algorithms are based on the heuristic assumptions that the best solutions will be found in regions of the parameter space containing a relatively high proportion of good solutions and that these regions can be explored by the genetic operators of selection, crossover, and mutation. Genetic algorithms offer a number of advantages: they search from a set of designs and not from a single design; they are not derivative-based; they work with discrete and continuous parameters; and they explore and exploit the parameter space (Goldberg 1989). The major disadvantage of this strategy is the computational cost of the large number of runs of the design code needed to evaluate a set of designs for each generation.

INTERDIGITATION, A HYBRID OPTIMIZATION TECHNIQUE

By using explicit, domain knowledge about the parameter space, the interdigitation technique achieves a significant gain in fewer runs of the code than standalone, domain-independent, numerical optimization techniques. Interdigitation of expert systems and numerical optimization focuses the search in areas that the engineer knows will be most beneficial. Since the engineer's knowledge of the domain is typically incomplete, interdigitation of numerical optimization and genetic algorithms is used to supplement the engineer's knowledge.

Figure 20.1 shows how interdigitation handles various degrees of knowledge about the parameter space. Knowledge is defined in terms of "causal relations" *and* "process." Causal relations specify regions in the parameter space and the design variables to change in order to get a desired movement in an output parameter either to escape a constraint or to increase the optimization value. Process specifies the order, direction, and magnitude to change the design variables specified in a causal relation. Figure 20.1a depicts an ideal parameter space, where the engineer has complete knowledge of the causal relations and the process. In this case, the interdigitation technique would be limited to expert systems since the engineer knows the exact steps to take in order to get to an optimal design.

Both Figures 20.1b and 20.1c represent cases where the engineer has incomplete knowledge of the parameter space. Figure 20.1b represents the case where the engineer knows all the causal relations but not all the process. Interdigitation combines expert systems with numerical optimization to solve this problem. The expert system identifies the causal relations, and numerical optimization determines the process. In Figure 20.1c the engineer has only partial knowledge of the causal relations and the process. Interdigitation combines expert systems, numerical optimization, and genetic algorithms. Expert systems and numerical optimization are used to pursue the engineer's knowledge; then numerical optimization and genetic algorithms are used to supplement it. Figure 20.1d represents the case where the engineer has no knowledge of the parameter space. Interdigitation combines numerical optimization and genetic algorithms to solve this problem. Detailed descriptions of the way interdigitation is used to pursue knowledge (domain-dependent) and to supplement knowledge (domain-independent) are presented in the following subsections.

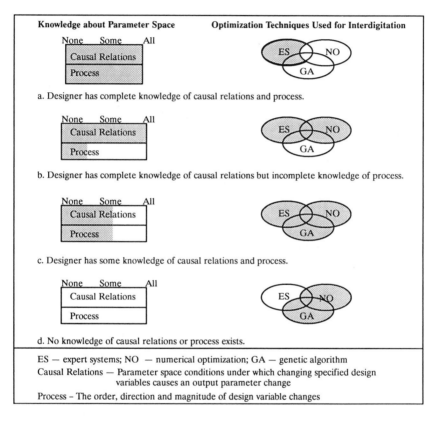

Figure 20.1: The interdigitation approach to optimization.

Domain-Dependent Search by Interdigitation of an Expert System and Numerical Optimization

The interdigitized technique captures the design engineer's explicit, domain knowledge in the form of rules. An intuitive menu-based interface lets the engineer add knowledge to the system without having to learn rule syntax, obviating the need for a knowledge engineer to extract information from the design engineer. The expert system uses the knowledge provided in the form of rules to focus on the key input parameters to be varied. Instead of varying all the parameters, those most likely to have an impact are varied first. This procedure reduces the number of design variables and the expenditure of computer time in determining gradient approximations for parameters that have little or no effect. For example, if a computer program has 100 design variables and the engineer knows that only 5 of them are important for the current location in the parameter space, the search is focused on those 5 variables. This focus reduces from 101 to 6 the number of runs of code needed to determine a search direction. The expert system searches the rule base for applicable rules and returns recommendations for design variables to be changed; it may also return information on the sequence, amount, and direction of change. If the process—direction and amount of change—is not specified in a rule, a one-dimensional or a multidimensional search technique is used to determine the amount by which a parameter should be changed.

The one-dimensional, direct search technique varies each parameter individually by an increment that is doubled or halved depending on the success of that change in improving the optimization value. The goal of this technique is not the continuous halving of a parameter increment until a tight convergence has been obtained. Rather, its goal is to reach an optimum value quickly for an individual design variable and then change to another design variable. Since the direct search approach is not based on gradient calculations, it is effective in discontinuous as well as gradient-insensitive regions.

The multidimensional, numerical optimization technique does not vary one parameter at a time but rather attempts to determine a single, composite search direction for all the design variables for which the direction and amount of change are not specified. This search direction is determined by finite difference techniques and is pursued until specified convergence criteria are met.

Interdigitation Rule Format.   The rule format was developed to allow the parameter space to be described in terms of causal relations and

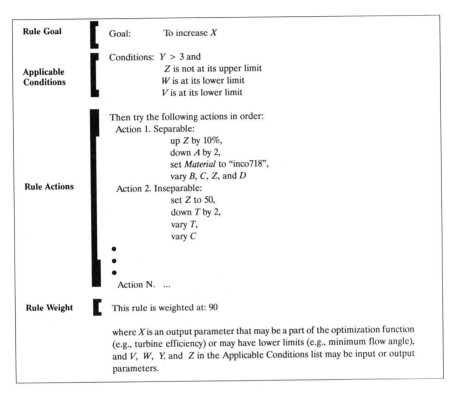

Figure 20.2: A sample rule.

processes. The four major components of a rule—Rule Goal, Applicable Conditions, Rule Actions, and Rule Weight—are shown in the sample rule in Figure 20.2.

Rule Goal and Applicable Conditions: The Rule Goal specifies an output of a design code and the direction in which it should be moved. For example, the goal of the sample rule in Figure 20.2 is to increase X; that is, the output parameter X needs to be increased. The Applicable Conditions under which a rule is valid consist of 0 or more expressions. If no expressions are specified, the rule is always applicable. To narrow the applicability of the rule, the engineer can supply information in the form of LISP expressions and symbolic constraints. LISP expressions allow the engineer to specify parameter space conditions exactly. For example, the sample rule specifies the condition that parameter Y must be greater than 3. Symbolic constraints are used for information about whether a parameter is at, above, or below a lower or upper limit but do not specify the actual numerical value of the limit. For example, the sample rule specifies that parameter Z is not at its upper limit and parameters W and V are at their lower limits, but it does not specify the values of those limits.

Rule Actions: The Rule Actions specify which input parameters to vary and how to vary them when the Rule Goal is the goal of the current optimization and the Applicable Conditions are satisfied. The Rule Actions are made up of "action operators" and "action types." The allowable action operators are: up/down, up-by/down-by, up-by-percent/down-by-percent, set, and vary. The up/down operators are specified when the engineer knows the direction in which to move a parameter but not the amount of the change. The amount is determined by the optimization technique being used. For example, if numerical optimization is used, the amount can be determined by a golden section method. The up-by/down-by operators are used when the engineer knows the direction of change and the amount by which the parameter should be varied. This amount is specified with a LISP expression which, when evaluated, determines the amount of change. For example, the action "down A by 2" in the sample rule would decrease A by an amount specified by the LISP expression that evaluates to 2. The up-by-percent/down-by-percent operators are similar to the up-by/down-by operators except that the LISP expression evaluates the percentage by which to change the parameter. For example, the action "up Z by 10%" would increase the value of Z by 10 percent. The set operator is used to set a parameter to an exact value. A LISP expression is evaluated, and the parameter is set equal to the value determined from the expression evaluation. The last operator is the vary operator. The engineer knows that a parameter is important but does not know in which direction to vary it. The optimization technique calculates gradient information to determine the direction of change.

The Rule Actions may specify many parameters to vary, but whether any of them are varied is initially determined by the rule type of Inseparable or Separable. Inseparable means that all the parameters specified to be varied are noninhibited parameters, that is parameters that are allowable design variables for this design *and* can be moved in the specified direction. In Action 2 of the sample rule, the action "down T by 2" is not possible if T is already at its lower limit. In that case, since the rule is inseparable, none of Action 2 is tried. This rule type is used when it makes sense to try an action only if all the specified parameters can be varied simultaneously. A Separable rule type removes inhibited parameters. The remaining uninhibited parameters make up the simultaneous rule action. In Action 1 of the sample rule, if the value of A is at a lower limit, then the parameter A is inhibited and the action "down A by 2" is eliminated as one of the rule actions.

Rule Weight: The Rule Weight is used to determine which rule to try first when more than one rule is applicable. The weighting allows the engineer to establish a priority ranking for the rules.

Interdigitation's Use of Rules. The interdigitation of expert systems and numerical optimization increases the effectiveness and eliminates the major weakness of each of these techniques. It enables the expert system to use knowledge that is incomplete. The rules do not have to state explicitly how much to vary each parameter. That is determined by the numerical optimization technique. On the other hand, the numerical optimization technique can take advantage of domain knowledge from the expert system and concentrate its search in a narrow region of the parameter space. Since the expert system uses rules that can have multiple operators for the same parameters, numerical optimization techniques can avoid local optima, constraint boundaries, and gradient-insensitive regions. For example, the interdigitized technique would use the actions of Action 2 of the sample rule—set Z to 50, down T by 2, vary T, vary C—to set Z to 50, decrease T by 2 and then calculate gradients for T and C from this new point. If the jump to the new values of T and Z and the subsequent movement of T and C by the numerical optimization technique does not increase the optimization value within a limited number of iterations, the previous design state is restored.

Domain-Independent Search by Interdigitation of Numerical Optimization and Genetic Algorithms

Many engineers have substantial experience in a particular domain. Since their knowledge is incomplete, however, domain-independent optimization techniques are needed to supplement their explicit knowledge. The interdigitized technique uses numerical optimization and genetic algorithms for this purpose. Unlike domain-dependent search, where the interdigitation of expert systems and numerical optimization uses only design variables that appear in rules, in domain-independent search, numerical optimization uses all the allowable design variables. For example, if a computer program has 100 design variables, all 100 are used by numerical optimization. The intent is to quickly climb to the local optimum. Exploration is sacrificed for the exploitation of numerical optimization to get the greatest gain in optimization value in the smallest number of runs of the code.

Genetic algorithms are used for exploration. The genetic algorithm is entered only after the "more efficient" exploitation techniques have been used. The genetic algorithm provides a robust fail-safe technique to avoid the traps that limit the progress of exploitation techniques, for example, a local optimum, a gradient-insensitive region, constraint boundaries, discontinuities, and input parameters that are not independent.

Implicit knowledge is gained about the parameter space during the exploration and exploitation of the parameter space by the domain-dependent search. Genetic algorithms provide a means to exploit that knowledge by using the states visited to seed the initial genetic algorithm population. This seeding provides "fertile" schemata for propagation and combination within the genetic algorithm. These schemata would eventually be discovered without seeding the genetic algorithm but only after thousands of runs of the design code. To enhance the efficiency of the genetic algorithm, a database of all previous designs is maintained so that a genetic algorithm does not rerun a design code that it has already run for this design.

One disadvantage of a standalone genetic algorithm is the high computational cost. Interdigitation reduces this cost by entering the genetic algorithm only after the more efficient exploration techniques of numerical optimization have been exhausted and by running the genetic algorithm only until an improvement in the fitness function is found. Once a better design is found, the interdigitized technique pursues this better design with the numerical optimization techniques. When the numerical techniques again arrive at an impasse, the genetic algorithm is reseeded with the best designs discovered by the numerical optimization and the past best designs from the previous genetic algorithm population.

Optimization Plan

Interdigitation is used for domain-dependent search, domain-independent search, and a combination of the two. Interdigitation couples a number of optimization techniques, switching from one to another within a single design run. The techniques to be used, their internal parameters, the run order of the techniques, and the way they are to be integrated are specified in an optimization plan. We have developed a default optimization plan that is very robust at obtaining an optimal design for a wide variety of parameter spaces. This default plan, shown in Figure 20.3, frees the engineer from having to know anything about optimization or having to tune the optimization technique to the engineering problem. This plan uses

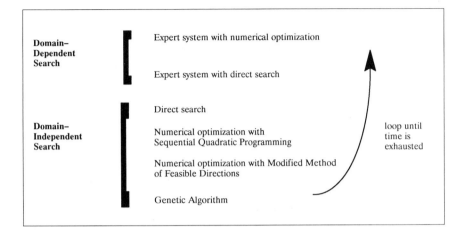

Figure 20.3: A default optimization plan.

domain-dependent search followed by domain-independent search. After the optimization plan is run, a transition is made back to the domain-dependent search techniques and the plan is rerun. The intent is to achieve the greatest gain in optimization in as few runs as possible. If time permits, however, exploration of the parameter space will continue. The transition from the genetic algorithm back to the other domain-dependent search techniques and domain-independent numerical optimization search techniques indirectly creates a genetic algorithm exploitation operator. The best design from the genetic algorithm is used as a new starting design for the next run of the optimization plan. Whenever the genetic algorithm is entered on subsequent runs of the optimization plan, the genetic algorithm population is reseeded. The reseeding involves replacing the worst members of the population with the better designs found, if any, from the exploitation techniques.

Our default plan can be modified—or a new plan can be tailored—for a particular parameter space. For example, an engineer may have a problem suited for numerical optimization and may specify an optimization plan that consists of two numerical optimization techniques: exterior penalty method followed by the modified method of feasible directions.

APPLICATIONS

The robustness of the interdigitized technique is shown by its performance on a test set of difficult engineering problems and on the design of a commercial airline turbine. The default optimization plan shown in Figure 20.3 was used for the test set problems and the turbine design.

Test Set of Engineering Problems

The six engineering problems listed in Table 20.1 were selected as representative of various types of parameter spaces that a single optimization technique must be able to handle. All six problems were in the list originally proposed by Sandgren (Sandgren 1977). Four of them—Chemical Reactor Design, Gear Ratio Selection, Five-Stage Membrane Separation, and Lathe Design—have not been included in numerical optimization studies because they have multiple optima and require large amounts of computer time (Gabriele 1988). Those features are exactly the reason for including them in our study. Real-world engineering problems are very complex, and the time required for a single run of a program ranges from seconds to a couple of hours for a finite element analysis. In Sandgren's study of nonlinear programming algorithms and in subsequent optimization studies, no single optimization technique solved all four problems. In fact, the best that any of Sandgren's 25 optimization algorithms could do was to solve two of them.

The design problems in our test set include the following features:

1. Variation of design variables, N, from 3 for Problem 2 to 16 for Problem 4.

2. Inclusion of inequality constraints, J, and equality constraints, K.

3. Initial starting points that are feasible (Problems 1, 2, 3, 6) and infeasible (Problems 4, 5).

4. Continuous and discontinuous parameter spaces (Problem 3).

5. Large and small feasible regions. E.g., Problem 4 has a very small feasible region where just finding a feasible solution is difficult.

6. Well-scaled and poorly scaled problems. E.g., design variables in Problem 4 range from 1E−6 to 1E+3.

The interdigitized technique was used under the following controlled conditions:

Problem	N	J	K	Feasible Starting Point	Value at Starting Point	Optimal Value Achieved		Active Output Constraints J
1. Process Design MAX Sandgren #3	5	6	0	Yes	3.0373	3.066 3.066 3.055 3.066	Sandgren MMFD SQP interdigitized	2
2. Chemical Reactor Design MAX Sandgren #9	3	9	0	Yes	−0.8756	−4.244 (−4.15114) −1.3 −1.3 −3.19	Sandgren SUN 4/330 MMFD SQP interdigitized	0
3. Gear Ratio Selection MIN Sandgren #13	5	4	0	Yes	0.2802	0.2679 0.2737 0.2737 0.2678	Sandgren MMFD SQP interdigitized	0
4. Five–Stage Membrane Separation MIN Sandgren #22	16	19	0	No	284.739	174.786 294.24 181.01	Sandgren MMFD interdigitized	16 (3 violated) 15 (0 violated)
5. Geometric Programming MAX Sandgren #23	7	4	0	No	2205.86	1809.76 1822.21 1820.01	Sandgren MMFD interdigitized	2 (1 violated)
6. Lathe MIN Sandgren #29	10	14	1	Yes	2931.0	−1614.938 +1300.6 +1300.5 −996.2	Sandgren MMFD SQP interdigitized	3 (0 violated)

N – Design variables   MMFD – Modified Method of Feasible Directions

J – Inequality constraints   SQP – Sequential Quadratic Programming

K – Equality constraints   Sandgren – Optimal value as reported by Sandgren

Table 20.1: Test set of engineering problems

1. No violation of constraints was allowed for a feasible solution.

2. All the engineering problems were run in double precision. No modifications to the code were made to obtain better results.

3. All the optimization was done on a SUN 4/330.

4. The numerical optimization algorithms used single precision and forward difference for all gradient calculations.

5. The *elitist* strategy was used for the genetic algorithm along with a standard crossover rate of 60 percent and a mutation rate of 0.1 percent.

6. The default optimization plan shown in Figure 20.3 was used for all the test problems. No settings were changed to enhance the performance of the interdigitized technique for an individual problem.

7. Since no knowledge was available from design engineers for these engineering problems, only the domain-independent search techniques of the interdigitized technique were tested.

The test results shown in Table 20.2 indicate the robustness of the interdigitized technique. The column labeled Optimal Value Achieved lists results from Sandgren (Sandgren 1977), the numerical optimization techniques of modified method of feasible directions , MMFD (Vanderplaats 1985) and sequential quadratic programming , SQP (Vanderplaats 1985), and the interdigitized technique. Since the default optimization plan uses MMFD and SQP, the results of the interdigitized technique will always be at least as good as the values shown for SQP and MMFD. On Problems 1 and 5, the numerical optimization technique was sufficient to obtain Sandgren's optimum. However, the interdigitized technique needed the combination of genetic algorithms and numerical optimization to achieve Sandgren's optimum on Problems 3 and 4 and to come very close to his optimum on Problems 2 and 6. On Problems 2, 3, 4, and 6, the numerical optimization techniques made progress but could not get to the optimum. For instance, on Problem 2, the numerical optimization could only get to -1.3, but with the combination of genetic algorithms and numerical optimization, the optimization could get to -3.19. Since the engineer does not know which optimization algorithm is best for the engineering design problem, the interdigitized approach offers a robust, general-purpose strategy for obtaining an optimal design.

On Problems 2 and 6, the final solution from the interdigitized technique varied significantly from Sandgren's optimum. The possible reason for this variance is that the interdigitized technique uses single precision and Sandgren's results are based on the use of double precision. We used single precision because the ADS numerical optimization package (Vanderplaats 1985) is written in single precision. The sensitivity of Problem 2 to the precision of its inputs is easily shown by analyzing the variation in the problem output when run at double and single precision. The result of running Problem 2 in double precision at the optimum input vector of (7828.79 188.81, 113.81) was −4.15, which is significantly different from Sandgren's result of −4.24. When this design was run at single precision for the optimum input vector, the result was −4.28 but a constraint was significantly violated, with a value of −3.82 instead of 0.47. The main conclusion to be drawn is that for these two problems the interdigitized technique gave significantly better results than a single numerical optimization technique.

Aircraft Engine Turbine Preliminary Design

The design of a high-pass turbine is a current real-world application at GE Aircraft Engine. The turbine consists of multiple stages of stationary and rotating blade rows inside a cylindrical duct. In the so-called preliminary design stage, all the critical dimensions (e.g., the inner and outer wall) and flow characteristics are determined without defining the detailed blade shape. For turbine design there are approximately 110 input parameters and 20 relevant output parameters per stage. The code used to model the turbine consists of approximately 10,000 lines of Fortran code, employs single precision, and takes approximately 30 seconds to run on a SUN 3/260.

Table 20.2 shows the results for an arbitrarily defined two-stage turbine design. The default optimization plan (Figure 20.3) used for the test problems was also used for the turbine design. In this case, however, explicit knowledge was available. An experienced engineer defined 16 rules involving eight design variables. The interdigitized technique outperformed numerical optimization by 60 percent. The advantages of interdigitation are shown in Figure 20.4. This bar chart compares the results listed in Table 20.2, the results obtained by the engineer not using any optimization tool, and the results obtained from running an expert system with directed search.

The final turbine was an unconventional design. The work per stage had a radically different distribution from traditional designs. Analysis of the optimization history shows that although the use of a genetic algorithm resulted in only a small gain in efficiency, it does appear to have pushed the optimization process away from being trapped in constraint boundaries so

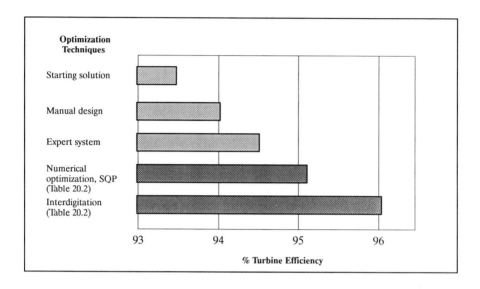

Figure 20.4: Optimization results for a two-stage aircraft engine turbine design.

Problem	N	J	K	Feasible Starting Point	Value at Starting Point	Optimal Value Achieved		Active Output Constraints J	Active Side Constraints
2-stage turbine	20	18	0	Yes	93.45	95.1	SQP	1	8
MAX						96.1	interdigitized	2	12

N — Design variables   SQP — Sequential Quadratic Programming
J — Inequality constraints   Active Side Constraints – Design variables at a lower or upper limit
K — Equality constraints

Table 20.2: GE turbine design.

that the local hill-climbing process could continue. In addition, the use of the interdigitized technique found one more active output constraint and four more active side constraints than numerical optimization.

CONCLUSIONS/FUTURE WORK

No single optimization technique exploits domain knowledge and also works for all parameter spaces. A hybrid, interdigitized technique combining expert systems, numerical optimization, and genetic algorithms provides an answer to the engineer's design problems. This hybrid technique runs the engineer's simulation codes unmodified and allows the engineer to concentrate on the design problem instead of the optimization algorithm. The genetic algorithm provides a fail-safe technique that avoids parameter space traps. To make the genetic algorithm an efficient optimization technique, it is used after exploitation techniques; it is seeded with good designs; and it uses a database to preclude running the same design more than once.

A test set of six problems showed the robustness of the interdigitized technique. The test set consisted of engineering problems with various constraints. This test set is recommended as a standard for the genetic algorithm and design automation communities. The interdigitized technique also showed a significant benefit over individual numerical optimization techniques when applied to the preliminary design of a two-stage turbine at GE.

Learning Domain-Dependent Rules

Future work on the interdigitized technique is focused on automatically learning domain-dependent rules and improving control of the genetic algorithm. The amount by which the expert system can increase the efficiency of numerical optimization and genetic algorithms is dependent on the amount and quality of the knowledge available about the parameter space. If the domain-specific knowledge is incomplete or inaccurate, the domain-independent search techniques will discover better designs. A major disadvantage of using numerical optimization and genetic algorithms in the absence of domain knowledge is the extra computational effort required for the large number of runs of the simulation code. We are investigating techniques of machine learning to analyze the improved designs achieved by domain-independent techniques in order to develop new rules that specify the parameters to vary and the conditions under which the rules are applicable. This additional knowledge will be used in future designs. It will make the optimization of subsequent designs more efficient and the rule base more nearly complete, and will provide the engineer with more knowledge about the parameter space.

Control over the Genetic Algorithm

At present, interdigitation uses the genetic algorithm only until a better design is found; then it switches to numerical optimization techniques for exploitation. One problem with the current implementation is that the genetic algorithm must not only find a better niche but also find a design with a better optimization value. We believe that the performance of the genetic algorithm might be greatly enhanced if all that was needed was to find a design from a better niche before applying hill-climbing techniques. We are currently investigating various approaches to niche identification.

ACKNOWLEDGMENTS

The authors would like to thank Garret Vanderplaats for providing his ADS numerical optimization package and John Grefenstette for providing his GENESIS genetic algorithm (Grefenstette 1984).

REFERENCES

Adeli, H. (1986). Artificial intelligence in structural engineering. *Engineering Analysis* 3.

Booker, L. (1987). Improving search in genetic algorithms. *Genetic Algorithms and Simulated Annealing.* London: Pitman, Chap. 5.

Gabriele, G. (1988). The generalized reduced gradient method for engineering optimization. *Engineering Design.* Amsterdam: Elsevier Science Publishing Co., Chap. 2.5.

Gero, J. (1988). Development of a knowledge-based system for Structural Optimization. *Structural Optimization.* The Hague: Kluwer Academic Publishers.

Goldberg, D. E. (1989). *Genetic Algorithms in Search, Optimization and Learning.* Reading, Mass.: Addison-Wesley Publishing Co.

Grefenstette, J. (1984). GENESIS software package.

Holland, J. (1975). *Adaptation in Natural and Artificial Systems.* Ann Arbor: University of Michigan Press.

Jabri, M. (1987). Implementation of a knowledge base for interpreting and driving integrated circuit floorplanning algorithms. *AI in Engineering* 2.

Powell, D., M. Skolnick, and S. Tong (1989). EnGENEous; domain independent machine learning for design optimization. *Third International Conference on Genetic Algorithms*. Los Altos, Calif.: Morgan Kaufmann Publishers.

Quinlan, J. (1987). *Applications of expert systems*. Reading, Mass.: Addison-Wesley Publishing Co.

Sandgren, E. (1977). *The Utility of Nonlinear Programming Algorithms*. Ph.D. Thesis, Purdue University.

Tong, S. S. (1986). Coupling artificial intelligence and numerical computation for engineering design. AIAA-86-0242, AIAA 24th Aerospace Sciences Meeting, Reno, Nevada.

Vanderplaats, G. (1985). ADS—A Fortran program for automated design synthesis. Available from COSMIC Computer Software, The University of Georgia, 382 East Broad Street, Athens, Georgia 30602.

Zhou, H. (1987). *CSM: a Genetic Classifier System with Memory for Learning by Analogy*. Ph.D. Thesis, Vanderbilt University.

21

Schedule Optimization Using Genetic Algorithms

Gilbert Syswerda

INTRODUCTION

Scheduling and planning are difficult problems with a long and varied history in the areas of operations research and artificial intelligence, and they continue to be active areas of research. They are problems that must be tackled with a combination of search techniques and heuristics.

Scheduling is difficult for two reasons. First, it is a computationally complex problem, described in computer science terms as NP-complete. This means that search techniques that deterministically and exhaustively search the space of possibilities will probably fail because of time requirements. In addition, search techniques that use heuristics to prune the search space will not be guaranteed to find an optimal (or even good) solution.

Second, scheduling problems are often complicated by the details of a particular scheduling task. This is true of the laboratory scheduling problem we will be discussing, where considerations such as optimization of operation setup are important. Algorithmic consideration of these specific constraints must often be embodied in what amounts to a domain-specific expert system.

How this knowledge is embedded into an optimization algorithm is often very algorithm-specific. In our approach, which uses a genetic algorithm as the main search technique, the domain knowledge remains separate. This provides us with considerable flexibility in adapting our techniques to particular applications.

THE SITS LABORATORY SCHEDULING PROBLEM

The scheduling problem we will discuss in the chapter has been motivated by the System Integration Test Station (SITS) laboratory of the U.S. Navy. The SITS laboratory is located at the Point Mugu Naval Airbase, on the shore of the Pacific Ocean in California. The laboratory contains F-14 airframes, complete with cockpit controls, avionics, radar, and weapon control systems, embedded in a simulation environment that causes the systems to behave as if they are in a plane flying at a specified altitude and speed. All cockpit controls and indicators are active and respond as if the plane is flying. The radar environment is also simulated, causing the radar and weapons control systems to behave as if they are sensing other planes (and missiles) in the space around them. Numerous pieces of support equipment such as computers, radios, and recorders are available as well.

The lab is made available to developers of the F-14 jet fighters. Multiple users can use the lab at once, creating a scheduling problem. The resources to be scheduled are the F-14 frames, the flight and radar simulation environment generators, and numerous pieces of support equipment. Rescheduling is also an important aspect of this scheduling problem. Once a schedule is in place, it is treated only as a working guideline, since high-priority emergency jobs are assigned preemptively, cancellations occur, and equipment breaks.

Currently, the schedule is constructed manually, requiring skill and a great deal of knowledge about the lab. The following characteristics make the lab scheduling problem difficult:

- **Resource constraints.** More than one user can use the lab at once, resulting in resource constraints among the simultaneous users.

- **Time constraints.** There exist global time constraints, such as no work scheduled after 5:00 P.M., and user-specific constraints. For example, a user may not be available at particular times or may prefer a particular time.

- **Setup time.** Certain tasks require extensive setup, which can take from one to two hours. There is more than one kind of setup, and doing one undoes another. Since users in this case are not billed for any setup that may be required, optimization of setup becomes an important issue.

- **Priority.** Tasks requested by users are assigned a priority, ranging from preventive maintenance to peremptory emergency tasks. When not all requested tasks can be placed into the schedule, those with the lowest relative priority should be left out.

- **Precedence.** Certain tasks cannot be performed until other tasks have been completed. Under these conditions, a task cannot be scheduled unless its predecessors have been previously placed into the schedule.

In addition to clearly definable requirements, real-world scheduling problems often impose scheduling considerations that are difficult to codify, perhaps because they concern a special job or the particular personnel involved in the task. As a result, for a schedule to be considered satisfactory, we must take into account both the resource/time constraint considerations and the more transitory (but equally important) aspects of the scheduling operation.

We are currently engaged in building a scheduling system that addresses these concerns. Our approach is twofold: to provide an intelligent schedule editor that understands the constraints of the scheduling problem and guides the user in building a schedule manually; and to provide a schedule optimizer that can work within the editor to cover a range of actions including completely manual, partially automatic, and completely automatic schedule construction.

SCHEDULE OPTIMIZATION

Our scheduling problem is primarily one of placing tasks that require time and resources into a matrix of resources versus time. If a task resides in the schedule at some time t, another task cannot be scheduled at t unless the intersection of their required resources is empty. A completed schedule consists of a number of potentially overlapping tasks in temporal order. (See Figure 21.1 for a sample schedule.)

```
    RESOURCES                                 RESOURCES1
a 3 .a....aa...aa.......a.........    .ac.ccaac..aac..cc.cacc.......
b 1 ..b....b.b..b.bb..b.b.........    .ac.ccaac..aac..cc.cacc.......   T
c 3 ..c.cc..c....c..cc.c.cc.......    .ac.ccaac..aac..cc.cacc.......   I
d 3 ..d...d..d..d.......d.......d    ..b....b.b..b.bb..b.b.........   M
e 3 ....eee..e.......e........e..e    .fdfffd.fd..d.......fdfff.f.fd   E
f 1 .f.fff..f..........f.fff.f.f.    ..d...d..d..d........d......d
g 3 g.g....g....g.ggg...gg.gg....    ..d...d..d..d........d......d
h 2 h...h.h.h.hh.................    g.g.eeeg.e..g.ggge...gg.gge..e
i 1 ...i.....iii....i..i.......i    g.g.eeeg.e..g.ggge...gg.gge..e
j 1 ........................j....j    g.g.eeeg.e..g.ggge...gg.gge..e
k 1 ..kk.k.k.....kk......k........    ..............................
l 1 l...ll..l...l..l.ll...l..l..l.    ..............................
m 3 ....m.m...mm............m...m.    ..............................
n 2 ..............nn..nn........n.    ..............................
o 3 .oo.....o.....o..oo..o........    ..............................
p 1 ....p.....p.pp....p..p.p......    ..............................
q 3 ...q...q....q..q..q.q..q.q.qq.    ..............................
r 1 r........rr.r.....r.......r.    ..............................
s 3 ...ss...ss.....s.....s....s...    ..............................
t 1 t............t..tt..........    ..............................
```

Figure 21.1: A sample set of tasks, labeled *a* through *t*, is listed on the left, along with the required number of hours and the resources required. The matrix on the right is a partially completed schedule, with tasks *a* through *g* placed into the schedule on a first-come, first-served basis.

This view allows us to consider scheduling as an ordering or combinatorics problem. What fundamentally must be done is to place a list of tasks in a particular order. The job is complicated by the fact that not all orderings of the tasks are legal, since they lead to constraint violations.

To circumvent the problem of illegal orderings, we use a deterministic schedule builder that takes a particular task sequence and builds a legal schedule from it. To do this, the schedule builder takes the first task from the ordered list and places it into the schedule according to domain-specific heuristics. It then selects the next task and places it into the schedule, taking care to respect the constraints of the task already there. This process is continued for every task in the ordered list, and what emerges is a legal schedule for the given ordered list of tasks.

Given that we can produce a legal schedule for a particular ordering of the tasks, what remains to be done is to find a task order that results in a good schedule. For that job, we turn to genetic algorithms.

The approach outlined above has a significant advantage in that the optimizer itself, the genetic algorithm, need not be concerned with details of the particular scheduling task. It can be limited to using syntactic operators on ordered lists. Domain knowledge can still be incorporated into the overall schedule optimization task, however, in the form of the schedule builder and evaluator.

A SIMPLIFIED SCHEDULING PROBLEM

To demonstrate the use of genetic algorithms in optimization of a schedule, we will use a simplified version of the SITS scheduling problem. We will attempt to construct a schedule for 90 tasks, each requiring 1 to 3 hours and a set of up to 30 resources. The 90 tasks must be scheduled into a block of 40 hours. There will be only a single instance of each resource, so a conflict for any resource will be a constraint violation. We will add a soft constraint and specify that some of the tasks prefer not be be scheduled during the first 12 hours. Figure 21.1 shows a partial task list and a partially constructed schedule. All tasks will start on the hour and will use time in one-hour increments.

CONSTRUCTION OF A GA-BASED OPTIMIZER

A genetic algorithm has four components that must be designed when use of a genetic algorithm to solve a problem is contemplated. These components are the syntax of the chromosomes, the interpretation and evaluation of the chromosomes, and the set of operators to work on the chromosomes. Other decisions such as the method for selecting parents must also be made, but these considerations are usually less dependent on the particular problem to be solved.

Chromosome Syntax

Traditionally, chromosomes are simple binary vectors. This simple representation has an appeal, and the theoretical grounding of genetic algorithms is based on binary vectors and simple operators. However, we run into problems when we attempt to represent complicated problems using binary vectors. Consider, for example, how we might encode a tour for the Traveling Salesperson Problem (TSP). In this problem, a collection of cities resides on a plane, and we have to find the shortest tour that visits each city only once.

An immediate coding scheme that comes to mind is to assign each city a binary number and to let the chromosome be a concatenation of these binary numbers. However, simple operators such as mutation and crossover can easily create binary strings that call for visiting some cities more than once, and others not at all.

A better approach is to change the chromosome syntax to fit the problem. In the case of the TSP, the chromosomes can simply be permutations of the list of cities to be visited. Each chromosome is then always a legal solution to the problem. Operators are devised that change the order of cities in parents to produce children with new orderings. An important objection to this approach is that we have left theory behind and are venturing into uncharted territory. The excuse we will use is that experience has shown that using other representations and operators can work very well and seemingly work in the same way that binary vectors work in traditional genetic algorithms. We will simply forge ahead and hope that theory will eventually chart the territory within which we wander.

In choosing a chromosome representation for the SITS scheduler, we have two basic elements to choose from. The first is the list of tasks to be scheduled. This list is very much like the list of cities to be visited in the TSP problem. However, the list of cities for the TSP very directly encodes a solution to the problem: the permutation *is* a tour of the cities. With the scheduler, however, an ordering of the cities does not directly represent a schedule. As already discussed, we must do something with that ordering to produce a legal schedule.

An alternative to using a sequence of tasks is directly to use a schedule as a chromosome. This may seem like an overly cumbersome representation, necessitating complicated operators to work on them, but it has a decided advantage when dealing with complicated real-world problems like scheduling. For example, imagine we have a schedule that is pretty good except that a high-priority task has been left out. A simple greedy algorithm running over the schedule could find a place for the high-priority task by removing a low-priority task or two and replacing them with the high-priority task. These kinds of operators, when combined with more global search operators, can greatly increase the speed at which genetic algorithms arrive at good solutions.

Considering again the other kind of representation, a simple sequence of tasks, we are confronted with the problem of what we should change in the ordering of tasks to accomplish the goal of replacing a couple of low-priority tasks with a high-priority task. Unless we have an extremely simple and forthright chromosome interpreter, it will not be possible to invert the schedule-building routine so that the chromosome that encodes for a particular schedule can be derived from that schedule. This means that local fixes to a schedule cannot be done at the level of the chromosome, generally barring us from creating domain-specific hill-climbing operators.

In our case, the appeal of a clean and simple chromosome representation won over the more complicated one. The chromosome syntax we use for the scheduling problem is what was described above for the TSP, but instead of cities we will use orderings of tasks.

Chromosome Interpretation

Having made our choice concerning the representation, we now have the challenge of converting an ordering of tasks into a schedule. This is accomplished by the schedule-building portion of our optimizer. The schedule builder, unlike the genetic algorithm, completely understands the details of the scheduling task. The schedule builder can be simple or very complex, but it must satisfy one overall constraint: the schedules it produces must be legal schedules.

Should the schedule builder be simple and stupid, or complex and smart? The answer to that question depends on how much faith we place in the genetic algorithm portion of the system. If the schedule builder is not very clever, it will seldom construct very good schedules. However, that a particular order is not very good is crucial information for the genetic algorithm, since it uses the relative ranking of the chromosomes in deciding what to do next. If the schedule builder is clever and does many optimizations on the schedule it builds, it will to a certain extent fix what is wrong with a particular ordering and cause a better schedule to be built than was coded for by the chromosome. In our system, the schedule builder is fairly simple. We have taken the approach that local optimization can be done once the genetic algorithm has taken its best shot.

Chromosome Evaluation

The evaluation of chromosomes is a critical portion of the system. The genetic algorithm uses the values generated to control parent selection and deletion. Furthermore, the evaluation function must capture what it is about schedules that makes them seem good or bad to the system's users. This is not always easy to do, since conflicting factors must often be given relative value, in addition to vaguely described preferences that are difficult to codify.

For our trial problem, we use a fairly simple evaluation function that captures concerns about task priority and time preferences. Since the schedule builder guarantees that no hard constraints are violated, the evaluation function needs to be concerned only with weak constraints and other user preferences (such as optimization of setup time). Hard constraints, such as time and resource requirements, are handled by the schedule builder and need not be evaluated.

For our sample problem, the evaluation of a schedule starts with the sum of the priorities of all the tasks. For each task not placed into the schedule, its priority is subtracted from the sum. For each task placed into the schedule, its priority is added to the sum unless it has a weak constraint violation, in which case only half its priority is added. With this scheme, if no tasks are scheduled, the evaluation function returns zero. If a perfect schedule is constructed, the evaluation function returns twice the priority sum.

Operators

One of the more interesting things to do when designing a genetic algorithm is to invent the operators that will construct the new potential solutions. Constructing operators can be tricky, and experimentation is often necessary to get them right. The chromosomes should be broken up in a way that is "natural" for the problem at hand.

For our scheduling problem, we have chromosomes consisting of lists of tasks that the schedule builder uses to construct schedules. What is important in an ordering of tasks that allows the schedule builder to build a good schedule? One thing that is clearly important, especially with regard to the greedy considerations of a single task, is the position of that task in the list. The closer the task is to the front of the list, the greater is its chance that it will be placed into the schedule.

Another somewhat more subtle consideration for a task is the nature of the tasks that precede it in the task list, since it is these tasks that will be imposing constraints on where it can be placed. If two tasks both require a scarce resource, the first task in the list may prevent the second from being scheduled, implying that the order of tasks is also important.

Random Search.   The first "operator" we will examine is the generation of random task lists; this will be our baseline for comparison. Given enough guesses, we could eventually guess the best task ordering. However, we are dealing with 90 tasks, making the size of the input space 90!, or about 10^{138}. This is far too large to search using exhaustive search or random guessing. Figure 21.2 presents the results of guessing 3000 times, while saving the best found thus far.[1]

The solutions generated by the random search are actually pretty good, since the schedule builder is interleaving the tasks presented to it rather effectively.

Mutation.   Mutation plays a dual role in genetic algorithms: it provides and maintains diversity in a population so that other operators can continue to work and it can work as a search operator in its own right. Mutation typically works with a single chromosome, and in our case, it always creates another chromosome (leaving the parent intact in the population). We will consider three mutation operators.

Position-based Mutation: Two tasks are selected at random, and the second task is placed before the first.

Order-based Mutation: Two tasks are selected at random, and their positions are interchanged.

Scramble Mutation: Supposing that the neighborhood of tasks in a task list is important, we choose a sublist randomly, and scramble the order of the tasks within the sublist.

---

[1]All data presented in this chapter are the average of 50 trials.

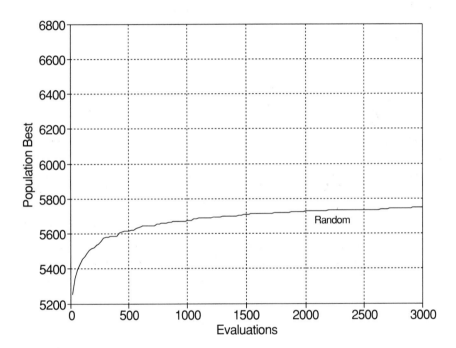

Figure 21.2: Random chromosome generation. The "operator" in this case was simply generating random task sequences and inserting them into the population.

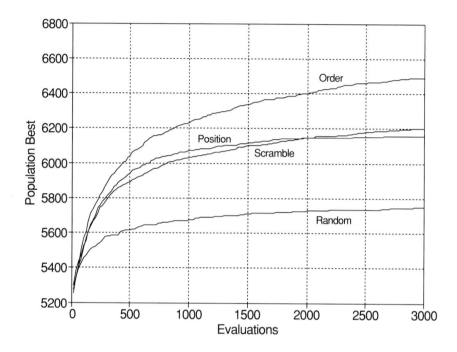

Figure 21.3: Mutation. Three mutation operators are presented and contrasted with simple random search. The order-based operator is clearly superior and in fact seems to be an effective search operator for this problem space.

How well do these operators work, compared with each other and compared with random search? Figure 21.3 presents all four. As we can see, these simple mutation operators are much more effective than simple random search. In addition, the order-based mutation operator is more effective than either the position-based or the scramble mutation operator.

Crossover. Crossover involves combining elements from two parent chromosomes into one or more child chromosomes. We will consider three crossover operators: one based on order, one on position, and the third invented by Darrell Whitley (1989) for problems of this sort. Each operator requires two parents.

Order-Based Crossover: A set of positions is randomly selected, and
the order of tasks in the selected positions in one parent is imposed
on the corresponding tasks in the other parent.

Parent 1:	a b c d e f g h i j
Parent 2:	e i b d f a j g c h
Selected positions:	* * * *
Child 1:	a i c d e b f h g j
Child 2:	b i c d f a j g e h

Position-Based Crossover: Again a set of positions is random selected,
but this time the *positions* of tasks selected in one parent are imposed
on the corresponding tasks in the other parent.

Parent 1:	a b c d e f g h i j
Parent 2:	e i b d f a j g c h
Selected positions:	* * * *
Child 1:	a i b c f d e g h j
Child 2:	i b c d e f a h j g

Edge Recombination Crossover: This operator involves building a table of adjacent tasks in each parent and then constructing a child using the adjacency information in the table. This operator, described
in detail in Whitley (1989), builds a child with tasks that are almost
always next to each other in one or the other parent. This operator,
if we use the naming style we have been using, might be called the
adjacency-based operator.

Results for all three operators are presented in Figure 21.4. Both the
position and order-based operators did well, quickly climbing to good performance levels. It is somewhat surprising that the edge recombination
operator did so poorly, given the success that Whitley has had with it. It
may be that adjacency is not an important property for this problem. This
makes sense given the way the schedule builder constructs schedules: it
always looks for the first available fit, and as a result tasks that are next
to each other in the task list often do not end up next to each other in the
schedule. Another possibility may be that the parameter settings used for
the genetic algorithm we were running were not optimal for this particular
operator.

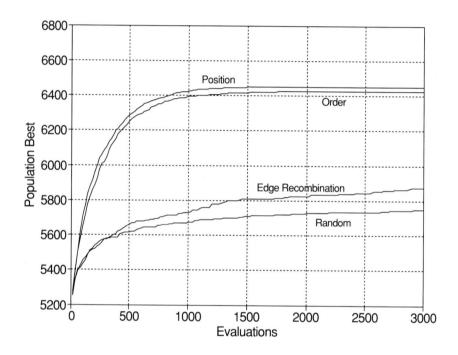

Figure 21.4: Crossover. The performance curves of the three crossover operators. For this problem, edge recombination does little better than random search.

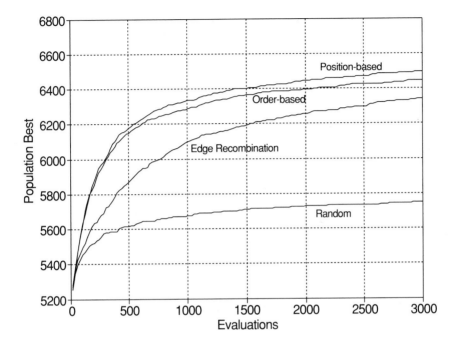

Figure 21.5: Operator Combinations. Each crossover operator was run in conjunction with the order-based mutation operator, each being applied at equal rates. The random search operator is presented simply as a reference.

Combining Operators. The order-based mutation operator is quite effective by itself, but it is slower than the order and position-based crossover operators. Moreover, combinations of operators can often support each other, giving rise to performance levels better than a single operator is capable of achieving. In light of this, order-based mutation was used in conjunction with each individual crossover operator. In each case, crossover and mutation had an equal chance of being applied. The results are presented in Figure 21.5.

The mix of operators could probably be improved by changing the relative use of each operator. Comparing the position-based crossover operator (Figure 21.4) and the order-based mutation operator (Figure 21.3), notice that crossover improves more rapidly than does mutation near the beginning of a run. However, after approximately 1000 evaluations, crossover stalls because the population has converged and there is nothing left for

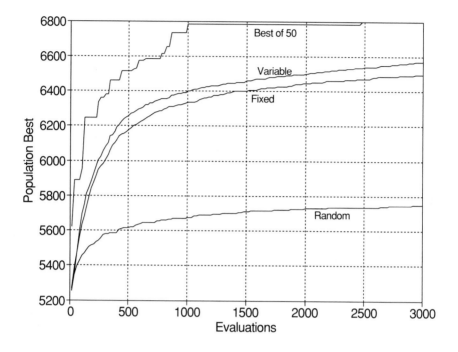

Figure 21.6: Performance of the genetic algorithm when slowly increasing the order-based mutation rate while decreasing the position-based crossover rate (Variable), compared with equal application of both operators (Fixed). Also presented is the best solution found of all 50 runs of Variable (Best of 50).

crossover to work with. Mutation on the other hand introduces diversity each time it is applied, and as a result the population continues slowly to improve. It would seem to make sense to apply crossover at increased levels at the beginning of a run and mutation at the end. Figure 21.6 presents the results of using this procedure, contrasted to applying each operator with equal probability.

Lawrence Davis (1989), carrying this line of reasoning further, has devised an algorithm that measures the effectiveness of each operator at every stage of a run, dynamically adjusting the probability of each operator according to its current effectiveness.

Population Size.   When designing a genetic algorithm optimizer, we must eventually confront the matter of deciding what the population size must be. Given that we were only going to perform 3000 evaluations, testing indicated that a population size of 30 was effective for the position and order-based operators we have been examining.

General wisdom dictates that a larger population will work more slowly but will eventually achieve better solutions than a smaller population. Experience indicates, however, that this rule of thumb is not always true, and that the most effective population size is dependent on the problem being solved, the representation used, and the operators manipulating the representation. It may be that Whitley's edge recombination operator would have been more effective in a larger population. Currently, only experimentation will give us that answer. David Goldberg (1989) has studied population sizes in the framework of binary representations and standard operators.

If you have only a finite number of evaluations you can perform before an answer must be given, how should you allocate those trials? You could use a larger population and use all your available evaluations on a single run. Or, you could use a smaller population, which might converge more quickly, and run it several times. For difficult problems like scheduling, each run will likely produce a different answer, and one of those answers will be the best. Will the best of a number of short runs be better than the best of a single longer run? Again, this question will require experimentation.

The experimental results presented throughout this chapter are all the average of 50 runs. Figure 21.6 presents the results of using position-based crossover and order-based mutation with sliding probabilities. Also presented is the best value found from all 50 runs at each stage of the optimization. The best value found is clearly better than the average value, which suggests that multiple runs may be a good strategy for this problem.

SUMMARY

In this chapter, we took a scheduling problem and in a step-by-step manner constructed a GA-based optimizer for it. Along the way, issues concerning chromosome syntax, interpretation, and evaluation were discussed, and various operators for constructing new chromosomes were presented. Many other matters were only mentioned or ignored entirely (such as parallel genetic algorithms), some of which are covered in the tutorial and in other chapters of this book. It is hoped that the material presented here will be useful to others in applying this interesting technology.

REFERENCES

Cleveland, G. and S. Smith (1989). Using genetic algorithms to schedule flow shop releases. *Proceedings of the Third International Conference on Genetic Algorithms and Their Applications.* San Mateo, Calif.: Morgan Kaufmann.

Coombs, S., and L. Davis (1987). Genetic algorithms and communication link speed design: constraints and operators. *Proceedings of the Second International Conference on Genetic Algorithms and Their Applications.* Hillsdale, N.J.: Lawrence Erlbaum Associates.

Davis, L. (1989). Adapting operator probabilities in genetic algorithms. *Proceedings of the Third International Conference on Genetic Algorithms and Their Applications.* San Mateo, Calif.: Morgan Kaufmann.

Davis, L. (ed.), (1987). *Genetic Algorithms and Simulated Annealing.* A volume in the Pitman Series of Research Notes on Artificial Intelligence. London: Pitman.

Davis, L. and S. Coombs (1987). Genetic algorithms and communication link speed design: theoretical considerations. *Proceedings of the Second International Conference on Genetic Algorithms and Their Applications.* Hillsdale, N.J.: Lawrence Erlbaum Associates.

Davis, L. (1985). Job shop scheduling with genetic algorithms. *Proceedings of an International Conference on Genetic Algorithms and Their Applications.* Hillsdale, N.J.: Lawrence Erlbaum Associates.

Goldberg, D. E. (1989). *Genetic Algorithms in Search, Optimization and Machine Learning.* Reading, Mass.: Addison-Wesley Publishing Co.

Goldberg, D. E. (1989). Sizing populations for serial and parallel genetic algorithms. *Proceedings of the Third International Conference on Genetic Algorithms and Their Applications.* San Mateo, Calif.: Morgan Kaufmann.

Holland, J. H. (1975). *Adaptation in Natural and Artificial Systems.* Ann Arbor, Mich.: University of Michigan Press.

Holland, J. H., K. J. Holyoak, R. E. Nisbett, and P. R. Thagard (1987). *Induction: Processes of Inference, Learning, and Discovery.* Cambridge, Mass.: MIT Press.

Syswerda, G. (1989). Uniform crossover in genetic algorithms. *Proceedings of the Third International Conference on Genetic Algorithms and Their Applications.* San Mateo, Calif.: Morgan Kaufmann.

Whitley, D., T. Starkweather, and D. Fuquay (1989). Scheduling problems and the traveling salesman: the genetic edge recombination operator. *Proceed-

ings of the Third International Conference on Genetic Algorithms and Their Applications. San Mateo, Calif.: Morgan Kaufmann.

22

The Traveling Salesman and Sequence Scheduling: Quality Solutions Using Genetic Edge Recombination

Darrell Whitley, Timothy Starkweather, and Daniel Shaner

INTRODUCTION

Scheduling poses a difficult problem in numerous application areas. Realistic scheduling problems involve constraints that often cannot be precisely defined mathematically. As a result, traditional optimization methods are not always applicable for many scheduling and sequencing problems or are not as effective as we would like. Genetic algorithms not only offer a means of optimizing ill-structured problems, but also have the advantage of being a global search technique.

We have developed a genetic operator that generates high-quality solutions for sequencing or ordering problems; the performance of this operator is all the more remarkable because it does not require any heuristic or local optimization information. In fact, as is typical with genetic search, all that is required is that it be possible to obtain some overall evaluation of a sequence relative to other sequences. We refer to this operator as genetic edge recombination. To demonstrate the effectiveness of this operator on a widely known sequencing problem we have applied it to the Traveling Salesman Problem. This problem involves finding the shortest Hamiltonian path or cycle in a complete graph of n nodes. This problem is a classic example of an NP-hard problem; all known methods of finding an exact solution involve searching a solution space that grows exponentially with n, the number of nodes in the graph. In this chapter we report results for 30 and 105 city problems that are superior to our previously reported results on these problems (Whitley et al. 1989).

Although this operator produces excellent results on the Traveling Salesman Problem, the significance of the operator goes well beyond this particular problem. Because the operator only looks at the total evaluation of a sequence, it can be applied to any general sequencing problem. It therefore represents a general approach to sequencing scheduling which produces a global search and which, despite its general purpose nature, produces results that are competitive with other, more domain-specific methods as illustrated in the Traveling Salesman Problem.

We have built a prototype scheduler that is based on a production line scheduling problem encountered at a Hewlett-Packard manufacturing site in Fort Collins, Colorado. Our recent work with this scheduling system looks at various assumptions about the flow of jobs along the production line. A careful analysis of this problem has revealed both new strengths and weaknesses in the operator for specific types of problems. Specifically, we compare a strict First-In-First-Out (FIFO) ordering of jobs to another approach that allows reordering of jobs between machines using heuristic rules. Surprisingly, our data suggest that using the genetic algorithm alone on the more difficult, strict FIFO sequencing problem can produce competitive results with the hybrid algorithm that combines the genetic search with greedy scheduling heuristics and that allows reordering of jobs between machines. This result not only demonstrates the advantages of global search, but it also makes our results more useful for a wider variety of real-world scheduling applications.

We have also studied classes of problems that cause difficulty for genetic search. We show that for a problem with multiple solutions the genetic algorithm may converge to a competitive, yet suboptimal, solution. Specif-

ically, this can occur when two competitive but dissimilar parents reproduce. By dissimilar, we mean that they represent different solutions, which
when recombined produce disfunctional or uncompetitive offspring.

Finally, the results presented here are in part based on enhancements we
have made to our genetic algorithm implementation. Our algorithm, which
we refer to as GENITOR, differs from most standard genetic algorithm
implementations in that it employs one-at-a-time recombination where the
best strings always survive and it uses a rank-based method of allocating
reproductive trials (Whitley 1989, Whitley and Kauth 1988). We have
shown that a distributed version of this algorithm can actually produce
improved results over the serial algorithm. This is important, not only
because of the potential for improved optimization, but also because a
distributed implementation can make it possible to obtain results an order
of magnitude faster than would be possible with a serial search.

THE EDGE RECOMBINATION OPERATOR

The key to solving a problem such as the Traveling Salesman Problem using
genetic algorithms is to develop an encoding that allows recombination to
occur in a meaningful way.

Consider the following tours: [A B C D E F] and [B D C A E F]. If we
extract the "link" information in the tour, we have "ab bc cd de ef af" and
"bd dc ca ae ef bf" as parent tours. We assume that the tours are circular;
the journey finishes again at the initial city, thus forming a Hamiltonian
cycle. Also note that edge "ab" has exactly the same value as "ba." All
that is important is the value of the link, not its direction. Thus, an
ideal operator should construct an offspring tour by exclusively using links
present in the two parent structures. On average, these edges should reflect
the goodness of the parent structure. There is no random information that
might drive the search toward arbitrary links in the search space. Thus, an
operator that preserves edges will exploit a maximal amount of information
from the parent structures. Operators that break links introduce unwanted
mutation.

The edge recombination operator we have developed uses an "edge map"
to construct an offspring that inherits as much information as possible from
the parent structures. This edge map stores all the connections from the
two parents that lead into and out of a city. Since the distance is the same
between two cities whether one is coming or going, each city will have at
least two and at most four edge associations (two from each parent). An

example of an edge map, the operation of the recombination method, and an example of recombination are given in figure 22.1.

The problem that has normally occurred with edge recombination is that cities are often left without a continuing edge (Grefenstette 1987). Thus cities become isolated so that a new edge has to be introduced. Using an edge map we can choose "next cities" in such a way that those cities that currently have the fewest unused edges (and thus appear to have the highest probability of being isolated) have priority for being chosen next. Of course, there must be a connecting edge to a city before it is actually a candidate, but at any point in time there are up to three candidates (and up to four in the case of the initial city), and the candidate with the fewest edge connections is chosen. In the case of a tie the algorithm chooses randomly from among the candidate edges.

Note that the one edge that the edge recombination operator fails to enforce is the edge that goes from the final city to the initial city. On Hamiltonian paths, this is not an issue, but on Hamiltonian cycles it means that a limited amount of mutation is likely to occur (on at most one link in the offspring, i.e., at a rate of $1/n$ where n is the length of the tour). By allowing different cities to occupy the initial position in different genotypes, this mutation is distributed over the various links.

Our test over several random recombinations for problems of various sizes indicates that mutation rates are low, typically ranging from 1 to 5 per cent.

Edge Recombination and Schema Theory

The theory behind most genetic algorithm implementations assumes that recombination and selection change the sampling rate of hyperplanes in an n-dimensional hypercube, where n refers to the length of some explicit or underlying binary representation of the problem. It is possible to show that our recombination operator in fact manipulates an underlying binary encoding of the Traveling Salesman Problem that largely conforms to the existing theory. Although the purpose of this chapter is to motivate applications largely by way of example, having a mathematical foundation that allows us to account for results not only makes it easier to explain and justify the approach, but it also provides a theoretical base from which to extend and improve the method. We will not review the fundamental theorem of genetic algorithms (rather, see Goldberg 1989, Holland 1975, Schaffer 1987, Whitley and Starkweather 1990) but Figure 22.2 does demonstrate how the Traveling Salesman Problem can be represented by a binary en-

The edge map stores all the connections from the two parents that lead into and out of a city. Consider these tours: [A B C D E F] and [B D C A E F]. The edge map for these tours is as follows:

A has edges to : B F C E   D has edges to : C E B
B has edges to : A C D F   E has edges to : D F A
C has edges to : B D A   F has edges to : A E B

THE RECOMBINATION ALGORITHM

1. Choose the initial city from one of the two parent tours. (It can be chosen randomly or according to criteria outlined in step 4.) This is the "current city."

2. Remove all occurrences of the "current city" from the left-hand side of the edge map. (These can be found by referring to the edgelist for the current city.)

3. If the current city has entries in its edgelist go to step 4; otherwise, go to step 5.

4. Determine which of the cities in the edgelist of the current city has the fewest entries in its own edgelist. The city with the fewest entries becomes the "current city." Ties are broken randomly. Go to step 2.

5. If there are no remaining "unvisited" cities, then STOP. Otherwise, randomly choose an "unvisited" city and go to step 2.

AN EXAMPLE

1. The new "child" tour is initialized with one of the two initial cities from its parents. Initial cities A and B both have four edges; randomly choose B.

2. The edge list for B indicates the candidates for the next city are A C D and F. C, D and F all have two edges: the initial three minus B. A now has three edges and thus is not considered. Assume C is randomly chosen.

3. C now has edges to A and D. D is chosen next, since it has fewer edges.

4. D only has an edge to E, so E is chosen next.

5. E has edges to A and F, both of which have one edge left. Randomly choose A.

6. A must now go to F.

The resulting tour is [B C D E A F] and is composed entirely of edges taken from the two parents.

Figure 22.1: The edge mapped recombination operator

coding and how edge mapped recombination results in the recombination of the underlying binary encoding.

This analysis also reveals another source of mutation (one pointed out to us by David Schaffer). Notice that a mutation occurs in the very last bit position. This is due to "mutation by omission." The edge "ef" occurred in both parents but was not passed on to the offspring. If one parent visits cities [... H C D F ...] and the other parent visits cities [... J C D K ...] then there are two ways of recombining these edges: one that uses the common edge "cd" and one that does not. For example, an offspring containing the fragment [...J C D F...] recombines the two parents and uses the common edge "cd." However, an offspring can inherit the following sequence fragments [...H C J...K D F...] without introducing any edge other than those taken from the two parents. We describe this as mutation by omission. This kind of implicit mutation occurs only if a single common edge is present; if more than one contiguous common edge occurs between two parents, the mutation, if it occurs, will be an explicit one, since at least one common edge will be isolated if the recombination operator does not use all the common edges together.

Despite occasional explicit and implicit mutation, edge recombination is clearly manipulating schemata in this representation space. If the genetic algorithm actually used this binary representation, there would be no clue as to which recombinations result in legal tours. Nevertheless, this characterization of the problem allows us to relate the behavior of our operator to the existing theory.

Results for the Traveling Salesman Problem

Our experiments with the Traveling Salesman Problem involve the use of a serial implementation of GENITOR and a distributed version of GENITOR that allows swapping (or "migration") of individuals between subpopulations. The distributed algorithm uses 10 subpopulations, each of which is performing a typical genetic search except for the fact that copies of the best strings are passed to neighboring subpopulations at regular intervals. The motivation behind the distributed algorithm as well as a more thorough discussion of results for a variety of problems are presented elsewhere (see Whitley and Starkweather 1990).

On 30 city and 105 city Traveling Salesman Problems, the distributed genetic algorithm consistently finds good solutions. The best known solution for the 30 city problem is 420. (Actually, it is 420 if one rounds off the distance between edges before summing; it is 424 if we sum the real valued edges and then round.) There has been some confusion about this in the

The tour [A B C D E F] has the following edges: ab, bc, cd, de, and ef. All possible edges in this six city problem are:

ab	ac	ad	ae	af	bc	bd	be	bf	cd	ce	cf	de	df	ef

The Hamiltonian cycle [A B C D E F] can be represented as a binary mask that indicates which links are present. This is illustrated in the following table.

ab	ac	ad	ae	af	bc	bd	be	bf	cd	ce	cf	de	df	ef
1	0	0	0	1	1	0	0	0	1	0	0	1	0	1

Consider again the tours [A B C D E F] and [B D C A E F]. We now add this second tour as the second binary mask in the table.

ab	ac	ad	ae	af	bc	bd	be	bf	cd	ce	cf	de	df	ef
1	0	0	0	1	1	0	0	0	1	0	0	1	0	1
0	1	0	1	0	0	1	0	1	1	0	0	0	0	1

Now consider the following offspring: [B C D E A F]. We add the mask for this tour as the third binary string in our table.

ab	ac	ad	ae	af	bc	bd	be	bf	cd	ce	cf	de	df	ef
1	0	0	0	1	1	0	0	0	1	0	0	1	0	1
0	1	0	1	0	0	1	0	1	1	0	0	0	0	1
0	0	0	1	1	1	0	0	1	1	0	0	1	0	0

To simplify the binary mask, we remove all those bit positions where neither parent has an edge, and indicate how recombination occurred.

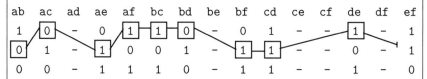

Notice that a mutation occurs in the very last bit position. This is due to "mutation by omission." The edge "ef" occurred in both parents but was not passed on to the offspring.

Figure 22.2: This figure illustrates how a Traveling Salesman Problem can be mapped to an underlying binary encoding and also indicates how edge mapped recombination acts on the underlying binary encoding.

literature. Oliver et al. (1987) report an optimal solution of 424, but the actual tour they give is not optimal: it has a length of 424 if we first round off the distance associated with the edges and then sum. This explains discrepancies we noted in a previous publication (Whitley et al. 1987) Using the serial version of GENITOR, a population of 2,000, and allowing up to 70,000 recombinations, we found the optimal solution of 420 on 28 out of 30 runs. On the other two runs we found the near-optimal solution of 421. We were somewhat surprised by these results. In our previous experiments we did not find solutions this consistently or this good. The only difference between these results and our previous results is a difference in population size. We had limited our experiments to populations of 100 to 500. Within this range, we found that populations of approximately 200 to 300 seem to work best. We went back and tested the larger population sizes only after obtaining the results from the distributed genetic algorithm. We ran the larger population sizes for comparative purposes, and thus were surprised by the results. From an applications point of view, this represents one of the traps that we must be careful to avoid. We originally did not push the genetic algorithm harder because we were already impressed by the results we had obtained.

For the distributed version of GENITOR we used 10 subpopulations of size 200 each, allowed up to 7,000 recombinations in each subpopulation, and swapped individuals between the subpopulations every 1,000 recombinations. The overall amount of work between the two algorithms is very comparable. The overhead of the distributed algorithm is roughly offset by the additional cost associated with sorting the larger population in the serial algorithm. Using the distributed algorithm, we found the optimal solution on 30 out of 30 runs. Thus, we were able to find the best known solution on every run.

On the 105 city problem we only present results for the distributed algorithm with migration. We used 10 subpopulations of size 1,000 each and allowed up to 200,000 recombinations in each subpopulation. Swapping occurred after every 10,000 recombinations. The best known solution (using the sum of the real-valued edges, and then rounding) is 14,383. We matched the best known solution on 15 out of 30 runs. This is 50 per cent of the runs; on the remaining 50 per cent of the runs, the solution found was always within 1 per cent of the best known.

SEQUENCING PROBLEMS

Our development of a theoretically well-founded genetic operator for the Traveling Salesman Problem is a major innovative step forward in the application of genetic algorithms to sequencing problems. The only information used was feedback about the total length of the tours that were evaluated. Although these results are impressive, the implications are far more dramatic for other kinds of sequencing problems where the problems are too ill structured to apply the kinds of heuristics traditionally used to solve Traveling Salesman Problems. Since the genetic recombination operator requires so little information, it is ideally suited to the many and varied kinds of sequencing problems that occur in scheduling.

The use of genetic algorithms lends itself to a general purpose system in the sense that the core of the genetic simulator remains the same from one application to the next. What varies from one application to the next is (1) the representation scheme and (2) the evaluation function. However, a broad class of scheduling problems can be viewed as sequencing problems. By optimizing the sequence of processes or events that are fed into a simple schedule builder, optimization across the entire problem domain can be achieved. Schedules have not always been viewed this way because there has not existed a general purpose mechanism for optimizing sequences that only requires feedback about the performance of a sample sequence. However, as our results on the Traveling Salesman Problem indicate, we have this capability. Thus, a "genetic" approach to scheduling has the potential to produce some very general scheduling techniques, and could be the foundation of a general purpose approach to sequence scheduling.

A Production Line Scheduler

We have developed a prototype system for optimizing scheduling problems. We have previously discussed a preliminary version of this system that used genetic recombination in conjunction with scheduling heuristics (Whitley at al. 1989). In this chapter, we show improved results on this type of problem as well as new results for a different version of the scheduling problem that uses no heuristics and enforces a strict FIFO ordering of jobs. We also look at variations on this same problem and clarify some of the assumptions of our model.

We have used the edge recombination operator to drive a prototype system designed to solve a production line scheduling problem modeled after a board assembly line at Hewlett-Packard in Fort Collins, Colorado. The problem is characterized by the following attributes:

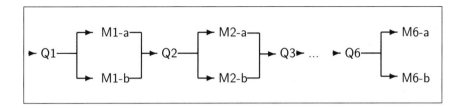

Figure 22.3: Layout of the production line.

- There are six workcells (i.e., machine groups) in sequence, and each completes a specific operation. Each workcell receives work from a single input queue and sends completed work to a single output queue. The scheduling dependencies are therefore six workcells deep. In order words, if we let A, B, C, D, E, and F represent the six workcells in sequence, then clearly the scheduling of workcell F is dependent on the scheduling of workcell E, which is dependent on D, which is dependent on C, which is dependent on B, which is dependent on the initial scheduling of workcell A. We stress the depth of these dependencies because the genetic algorithm must generate schedules that are sensitive to distant and indirect dependencies.

- Each workcell contains two identical machines operating independently and in parallel.

- Each machine has a "cost" associated with the type of work being done. In general, these are assumed to be the following: $1 per time unit for idling, $5 per time unit for processing, and $10 per time unit for setup.

- Twenty different types of products are produced in a single schedule. Each product has a fixed setup time and processing time at each operation.

Ten units of each type of product are to be assembled. Figure 22.3 illustrates the layout of the production line. M refers to Machine, and Q stands for Queue. Q1 is the initial job sequence optimized by the genetic algorithm. M1-a and M1-b are two identical machines that together make up the first workcell.

The operation of this hypothetical line has been modeled in two ways. In one case, a FIFO model is used such that a workcell simply dequeues the next item from its input queue and sets up for the job as required.

In the FIFO model any two adjacent workcells in the production sequence behave as two machines coupled by a conveyor belt. When a time unit is simulated, the next available job in the preceding queue is assigned to the next available machine. No attempt is made to give the next available job to a machine already set up for that machine. In the FIFO model, the only optimization is the optimization of the sequence in which jobs are supplied to the first workcell. In other words, only Q1 is optimized.

The HYBRID model involves optimizing the initial sequence of jobs (i.e., Q1) in conjunction with greedy heuristics for reordering jobs between workcells. Each workcell has random access to the jobs in the preceding queue as well as knowledge about the parallel machines and the "supplying" workcell. Whenever possible, the HYBRID model attempts to select a unit identical to the last and will wait idle if a "like unit" will become available before a different setup can be accomplished. This implies, for example, that machines M2-a and M2-b use locally available information to select jobs from Q2 in a different order from that in which they arrived in Q2.

The strategy behind the HYBRID/greedy genetic scheduler implementation is as follows. (1) Using the edge recombination operator, the genetic algorithm produces a "primary schedule" indicating the sequence in which jobs are loaded onto the first workcell. (2) The "secondary" rule-based scheduler then generates a full schedule. The rules for the secondary scheduler are relatively simple since the problem is highly constrained by the availability of jobs as they come off the first machines.

In both the FIFO and the HYBRID model, the evaluation of the resulting schedule is used to assign a fitness value to the initial sequence (i.e., Q1) which was generated by the genetic algorithm. This "fitness" value is used to rank the sequences and thus to allocate reproductive trials in a population of initial "job sequence" genotypes. Since the FIFO scheduler is not allowed to reorder jobs between machines, the FIFO scheduling task is more difficult.

For the HYBRID model, the solution space is highly constrained once the initial machine is scheduled. If a good sequence of jobs can be found, then the remaining machines can be scheduled using a few simple rules that act as a greedy scheduler with some lookahead capabilities. The greedy scheduler uses the initial job sequence to produce a complete schedule that is then evaluated. This evaluation is then assigned as the "performance" value associated with the initial sequence. Thus, the goal of recombination was to find a sequence that optimized the performance of the final schedule in the environment of the secondary greedy scheduler.

The rules used to implement the secondary scheduler of the HYBRID model do not actually look ahead but rather use a "wait and see" approach.

If a machine is already set up to run "Job X" and a job X is available, then it is scheduled. If a machine is set up to run "Job X" and X is unavailable, then an "idle time" is introduced. The scheduler will wait and see if this idle time can be filled later by another job. If X becomes available before another job can be set up, then the scheduler waits for X. However, if at some later point in time X is still not available and another Job Y could have been set up and started, then the idle time is "back-filled" by scheduling a setup for the new Job Y. If multiple jobs could be set up in place of X (the unavailable job the machine is currently set up to run), then further comparisons are made. First, the machine decides which jobs have the shortest setup time. If two or more jobs have identical setup times, the one that has the shorter execution time is chosen. The only restriction on the selection of the new job type is that if there is more than one type of job in the input queue, a machine will not set up to run the same job type being run by the corresponding parallel "like-machine." For example, the only way that M1-b will set up to run a job type that M2-a is already running is if that is the only job type in the input queue. This restriction overrides the "short-setup-time" criteria for selecting a new job type.

Both the FIFO and HYBRID models attempt to keep all machines busy at all times. In both models, the machines in the workcell will set up to do whatever work is available at almost every opportunity (subject to the restrictions just outlined). There is no attempt to determine if a machine should remain idle rather than set up or do work; machines are idle only if there is no work to do.

Another key part of adapting our genetic recombination operator for scheduling applications was to encode the problem as a sequencing problem. Using A, B, C and so on to represent jobs, we might have an actual sequence something like the following: [A A A B B A A C C]. Note that job types reoccur. Our strategy for encoding the problem was simply to ignore the fact that some of the jobs are the same and thus can change positions without altering the schedule. Thus, [A A A B B A A C C] is encoded as [A1 A2 A3 B1 B2 A4 A5 C1 C2]. This means we are not using all the information available to us. The results indicate that the operator works well even without this additional information. Although the results are positive, it is most likely possible to devise a version of the edge recombination operator that exploits this additional information. We have not vigorously explored this possibility, although this could perhaps solve some of the difficulties we discuss later in this chapter.

Workcell	Average Processing Time	Average Setup
1	5.20	3.40
2	6.20	5.55
3	4.70	3.90
4	4.70	3.90
5	4.70	3.90
6	4.70	3.90

Table 22.1: Scheduling Problem-1

The Scheduling Results

We evaluated our scheduler on numerous problems. We found that the most difficult of these problems involved schedules with a uniform number of jobs. While a uniform number of jobs is not typical of scheduling problems, we have found that this makes the problem more difficult. If there is a bias such that there are significantly more of one type of job than others, then the scheduling task is usually easier. This appears to make intuitive sense. If there are one or two jobs that occupy large blocks of time, then these jobs very much constrain the solution space. Large blocks of time can be allocated to the most frequently occurring job types, while any resulting "holes" are filled with small jobs.

The following discussion pertains to a problem we will refer to as problem-1. This problem has 200 job batches, where there are 10 different job types and 20 of each type. Setup times were varied from 1 to 10 time units, and processing times from 1 to 20 time units, as shown in Table 22.1. Workcells 3 through 6 have identical characteristics, while 1 and 2 vary significantly.

A theoretical limit can be calculated for problem-1 which establishes a lower bound on the cost function. The theoretical limit assumes that there is only one setup per job type per workcell, and no idle time. The theoretical limit for problem-1 has an associated cost of 35,110. If we assume a single line of machines is used rather than paired machines, the theoretical limit based on the assumption of one setup per job and no idle time does not change. Nevertheless, we can prove that in fact it is impossible to produce a schedule with a cost of less than 36,442 because no machine later in the sequence can finish before any prior machine in the sequence. The situation is more complex when paired machines are involved because we can now only make the following, less restrictive observation: at each workcell, at least one of the paired machines must finish after all preceding machines in the sequence are done. More than likely, the "real limit" has an associated cost of at least 36,400, and it may actually be somewhat above this amount.

Because this problem is fairly well structured, we were able to generate competitive schedules by hand. We did this as an iterative process, generating what appeared to be good schedules by hand and then evaluating them using the same schedule evaluation that the genetic algorithm uses. In this way, we were able to construct fairly tight schedules. For the HYBRID model we were able to develop a schedule with an associated cost of 40,363, and for the more difficult FIFO model we were able to construct a solution with an associated cost of 41,694. In all likelihood, the optimal solution for the HYBRID model has a lower cost than the optimal solution for the FIFO model, simply because job reordering between machines results in greater flexibility in the schedules we can generate. Moreover, a good HYBRID model schedule is not necessarily a good FIFO model schedule. The 40,363 HYBRID model schedule, when run through the FIFO scheduler, produces a schedule with an associated cost of 45,800.

Figures 22.4 and 22.5 show several different kinds of experiments. Two versions of the FIFO and two versions of the HYBRID algorithm were run. For each model, a set of experiments was conducted using a "single-population" genetic algorithm with a population of 50 and one using a distributed genetic algorithm with 7 subpopulations of 50. (We also ran serial populations of 350 for comparison; the end results were almost identical with the distributed genetic algorithm, although the distributed algorithm produced faster initial reductions in error, even after the results were rescaled for comparative purposes.) Figure 22.6 gives an overlay of the four sets of experiments with respect to the theoretical limit. Actually, in one sense the FIFO and HYBRID schedulers are solving two distinct versions of the scheduling problem, and so, in terms of the optimization results, we are comparing apples and oranges. On the other hand, from an applications perspective of the end result, we are ultimately solving the same scheduling task. Thus, from this perspective the comparison is an interesting one.

Figure 22.6 shows that both of the FIFO runs overtake the HYBRID algorithm after approximately 10,000 recombinations. In the serial runs with small population sizes, the HYBRID algorithm produces slightly better results than the FIFO in the end. However, when the more powerful distributed algorithm is used, the FIFO actually produces the best results. Furthermore, the comparisons to the "iterative/hand-generated" results are very interesting with respect to the two types of models. By using the HYBRID model, the distributed genetic algorithm produced an average result of 40,952 over 30 runs. Thus, its average behavior is just above the hand-generated solution of 40,363. However, using the distributed genetic algorithm to optimize the FIFO model we obtained an average result of

40,804, whereas our best hand-generated solution was 41,694.

Significance of the Scheduling Results

From a scheduling/application point of view, the results suggest that there is no significant difference between the FIFO and HYBRID scheduling systems. From an implementation and computational point of view, however, there are real differences. First, the FIFO model is computationally more efficient than the HYBRID model, since we no longer have the overhead of evaluating the small rule-base used to implement the greedy reordering heuristics. Second, the FIFO converges to competitive solutions in fewer recombinations. The FIFO runs are close to the final results by 15,000 recombinations, whereas the HYBRID runs do not catch up until after roughly 25,000 recombinations. The faster FIFO evaluation and convergence means that approximate results can be obtained two to three times faster than is possible with the HYBRID method.

The third difference, and certainly the most important one, is that FIFO is a more realistic model of many scheduling tasks. To implement the HYBRID algorithm would require human intervention or a sophisticated automated reordering system. The FIFO simply conveys work from one workcell to the next.

At first it may seem unreasonable that the FIFO scheduler should produce results competitive with the HYBRID system. However, we believe that the FIFO scheduler is competitive because it relies on global search. That is, what's good from a local, greedy point of view may not be good from the point of view of developing a good overall schedule. When tracing the FIFO schedules by hand we noted that the genetic algorithm appeared to miss optimization opportunities that the greedy heuristics would have certainly caught. However, when we traced the consequences of these "missed opportunities," we found that if the local improvement had been made it would have degraded the overall schedule because of dependencies that occurred later in the schedule.

The somewhat surprising results of the FIFO implementation is good news from our perspective because it suggests that we should not attempt to build heuristics into the scheduling system, but rather should let the genetic algorithm do the work. This keeps the implementation and the system simple and reduces computation costs.

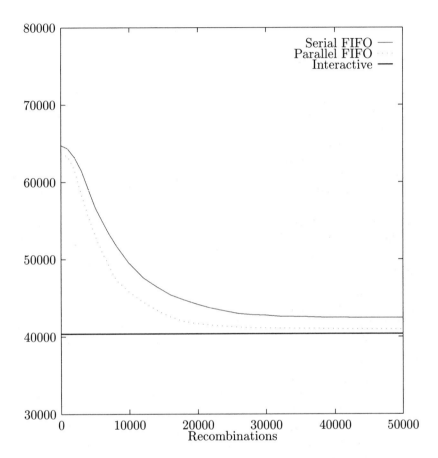

Figure 22.4: The results of using the HYBRID model in conjunction with a serial genetic algorithm (dashed line) and the distributed genetic algorithm (solid line). The interactively hand-generated solution (40,363) in this case is slightly better than the average solution found by the genetic algorithm. The average value of the solutions produced by the serial algorithm is 42,455, whereas the average value of the solutions produced by the distributed algorithm is 40,952. These results are averaged over 30 runs.

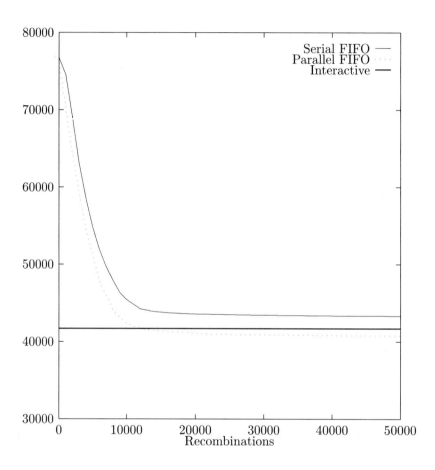

Figure 22.5: The results of using the FIFO model in conjunction with a serial genetic algorithm (dashed line) and the distributed genetic algorithm (solid line). In this case, the distributed FIFO genetic scheduler on average found solutions better than the interactively hand-generated solution (41,694) for this problem. The average value of the solutions produced by the serial algorithm is 43,330, whereas the average value of the solutions produced by the distributed algorithm is 40,804. These results are averaged over 30 runs.

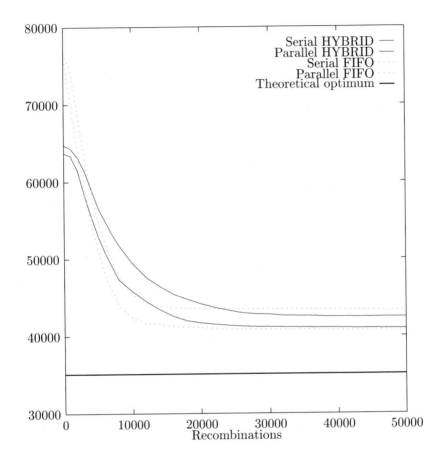

Figure 22.6: Overlay of the results of the FIFO model and the HYBRID model from the previous two figures. The two dotted lines represent the two FIFO results, whereas the two solid lines represent the HYBRID results. The theoretical optimum is also shown. In both cases the distributed solution is the result that is closer to the theoretical optimum.

ANALYZING THE SCHEDULER USING WORST CASE PROBLEMS

We have spent a good deal of time trying to decide how our evaluation function affects the schedules generated. Our analyses suggest that this problem is actually one that is relatively difficult to solve by genetic recombination. We have also conducted experiments studying the effect of using two machines in parallel. Our tests, though not exhaustive, suggest that running two machines in parallel while at the same time assuming that paired machines form a single workcell with a single queue between workcells makes the scheduling problem harder. In general, to obtain good schedules, the two parallel lines should specialize. That is, all "right-hand side" machines should do job A, while the "left-hand side" machines should do job B. However, to do this, the initial sequence of jobs must be perfectly interleaved to make the next available job arrive in the proper sequence to be consumed by the proper machine.

To use a simple case as an example, if we assume that every machine processes every job at the same rate and that all jobs have identical setup costs, then to achieve a tight schedule with minimal setups requires a sequence such as the following [A B A B A B .... C D C D C D]. In other words, this will cause the "right hand machine" to consume a job of type A, the "left-hand machine" to consume a job of type B, the "right-hand machine" next to consume a job of type A, and so on. Thus, a very simple scheduling task has a relatively complex (though highly regular) sequence of jobs. If we construct a problem of 200 jobs, with 20 different types of jobs and 10 jobs of each type, *and we assume that all processing times and all setup times for all jobs are the same,* then we create a situation in which there are numerous solutions. Furthermore, the multiple solutions for this schedule do not recombine well. Since we assume that all processing times and all setup times require five time steps, we will refer to this as the "All-5s" problem. A perfectly packed solution for the All-5s problem can be achieved which has a cost of 36,075. We stress that this is an artificial problem designed to create difficulties for the genetic algorithm. Nevertheless, it also provides useful insight into the nature of genetic search which impacts potential applications.

To illustrate why such a problem causes difficulties, consider a scaled-down version of the All-5s problem where we have four job types (A, B, C, and D) and we have two jobs of each type. All of the following sequences are perfect solutions to this problem: [A B A B C D C D], [A C A C B D B D], [C B C B A D A D], and so on. However, if we try to recombine some of these different potential solutions, we typically obtain uncompetitive off-spring. For example, we might recombine [A B A B C D C D] and [A C A C B D B D], both of which are optimal solutions. When recombining these "parents" it is possible to produce the offspring [A B D C A B C D], which is in fact the worst possible schedule. When dissimilar solutions are recombined and uncompetitive offspring are produced, the result from a theoretical point of view is high variance in the fitness values associated with hyperplane samples in the search space. This in turn means that the genetic algorithm is receiving inconsistent feedback about what are good sequence combinations (i.e., good hyperplanes) in the problem space. We have observed the same kind of difficulties in other problems (Whitley et al. 1990) while Baker and Grefenstette (1989) discuss from a theoretical perspective the problem of hyperplane samples that are characterized by a high variance in the associated set of fitness values for that hyperplane. Interestingly, the compromise solution to this problem is simply to block all jobs of the same type together; this compromise solution has an associated cost of 42,075. While there are multiple ways in which the jobs may be blocked, the "order" in which the blocked jobs occur is not important. Therefore, reordering the blocked jobs during recombination does not degrade performance in the offspring. The result of this strategy is that each machine in each workcell runs half of each job type. The schedule is perfectly packed, but two setups per job type are required.

Our past experience with other problems where it is possible to recombine dissimilar parents suggests that when this occurs we can obtain better results running several small searches instead of one larger search. This is because as larger populations are used, a greater number of dissimilar solutions are likely to be contained in the population.

We ran the All-5s problem using populations of 50 and a population of 500 using the paired machines configuration. On 30 runs the FIFO model using a population of 50 produced an average schedule of 39,923. Although we did not run a large number of experiments to confirm the following observation, based on runs using a population of 500 it appears that increasing the population size actually degrades performance toward the compromise solution. Again, this is not unexpected given that larger populations will contain more dissimilar solutions. The computational cost of running a population of 500 is actually comparable to running 10 populations of 50 for one-tenth the number of recombinations. Therefore, rather than increasing population size, if we take the best of 10 runs using a population of 50, we are much better off: an average of 37,628 is obtained in our experiments. Despite doing our best to torture the genetic algorithm with an extremely unfavorable problem, the results are respectable.

We next ran a set of experiments aimed at removing the obstacle that causes the difficulty for the genetic algorithm. We assume exactly the same All-5s problem, but we now assume that only a single line of machines is used. For the genetic algorithm the problem is much easier.

For a single line of machines (again assuming all jobs have the same processing time and setup costs), the sequence would be merely [A ... A, B ... B, etc.]. Although this is an easy problem (for people who happen to be good at finding order in complex patterns), the genetic algorithm must "discover" that this is an easy problem by finding that it should order the jobs in batches. GENITOR finds the the optimal solution within 10,000 recombinations on every run (30 runs each, FIFO and HYBRID) using populations of 50. This is not an amazing accomplishment, but it is reassuring that the genetic algorithm also finds this to be a trivial problem to solve.

DISCUSSION AND CONCLUSIONS

Our results on Traveling Salesman Problems are excellent, and, in general, we would not expect difficulties on problems sequencing unique items or events. For example, we are currently working on a real-world scheduling application where we must determine the sequence in which customer orders are to be filled. These orders involve different degrees of product mix (e.g., 50 units of product A, 20 units of product C, etc.) where there are approximately 500 products. The ability to fill an order is constrained by current production (not all products are being produced at any point in time) and by the products in storage (which we wish to minimize). Furthermore, in general we want to move the product from the production line into a truck for shipping. Therefore, once we start to fill an order, we wish to fill it as quickly as possible. Assuming we fill 10 orders per hour (which is realistic given the problem we are working on), then on average, we must fill and load each order for shipping in under an hour.

Although there are some subtle difficulties in solving such a problem, we view the underlying scheduling problem in much the same way that we view the Traveling Salesman Problem. Each order is unique, and we wish to find the sequence in which the orders should be filled. Our objective is to minimize the amount of time it takes to fill an order while at the same time minimizing the amount of product that must go in and out of storage. The exact balance to strike between these two goals will depend on the evaluation function.

In general, we feel that our scheduler will generate high-quality solutions to problems in which the objects or events to be sequenced are unique. What about scheduling problems in which the objects or events to be scheduled are not unique? Our experiments show that we can indeed create scheduling problems that are unfavorable for the genetic algorithm. We can even explain these problems in terms of the underlying theory of genetic algorithms. Nevertheless, unfavorable conditions do not prevent the genetic algorithm from finding competitive solutions, although they may prevent it from finding an optimal solution. Problems most likely to cause difficulty are those that have multiple dissimilar solutions that do not "recombine well." We suspect that the problems most likely to be difficult for the genetic algorithm may be those that are not likely to be hard problems in the first place. If a problem is sufficiently well structured to have multiple solutions that are essentially variations on the same scheduling strategy, then these solutions are likely to be obvious. Thus, in general, we do not believe that scheduling problems in which the objects or events to be scheduled are not unique pose a real difficulty.

Certainly, more work needs to be done. There is a critical need for comparative studies on larger problems that are known to be difficult and for which other scheduling strategies have been developed. However, the work conducted so far has produced positive results that suggest a very exciting potential.

Acknowledgments.   This research was supported in part by a grant from the Colorado Institute of Artificial Intelligence (CIAI). CIAI is sponsored in part by the Colorado Advanced Technology Institute (CATI), an agency of the state of Colorado. CATI promotes advanced technology education and research at universities in Colorado for the purpose of economic development.

REFERENCES

Baker, J. and J. Grefenstette (1989). How genetic algorithms work: a critical look at implicit parallelism. *Proceedings of the Third International Conference on Genetic Algorithms*, Palo Alto, Ca.: Morgan Kaufmann. pp. 20-27.

Goldberg, David (1989). *Genetic Algorithms in Search, Optimization, and Machine Learning.* Reading, Mass.: Addison-Wesley.

Grefenstette, John (1987). Incorporating problem specific knowledge in genetic algorithms. In Lawrence Davis (ed.), *Genetic Algorithms and Simulated An-*

nealing. Palo Alto, Ca.: Morgan Kaufmann. pp. 42-60.

Holland, John (1975). *Adaptation in Natural and Artificial Systems.* Ann Arbor: University of Michigan Press.

Oliver, I. M., D. J. Smith and J. R. C. Holland (1987). A study of permutation crossover operators on the traveling salesman problem. In John Grefenstette (ed.), *Genetic Algorithms and their Applications, Proceedings of the Second International Conference.* Hillsdale, N.J.: Erlbaum Associates. pp. 224-230.

Schaffer, David J. (1987). Some effects of selection procedures on hyperplane sampling by genetic algorithms. In Lawrence Davis (ed.), *Genetic Algorithms and Simulated Annealing.* Palo Alto, Ca.: Morgan Kaufmann. pp. 89-103.

Whitley, Darrell (1989). The GENITOR algorithm and selective pressure: why rank-based allocation of reproductive trials is best. *Proceedings of the Third International Conference on Genetic Algorithms,* Palo Alto, Ca.: Morgan Kaufmann. pp. 116-121.

Whitley, D. and J. Kauth (1988). GENITOR: a different genetic algorithm. *Proceedings of the Rocky Mountain Conference on Artificial Intelligence.* Denver, Colo. pp. 118-130.

Whitley, Darrell and Tim Starkweather (1990). GENITOR II: A distributed genetic algorithm. *Journal of Theoretical and Experimental Artificial Intelligence,* (in press).

Whitley, Darrell and Tim Starkweather and Chris Bogart (1990). Genetic algorithms and neural networks: optimizing connections and connectivity. *Parallel Computing,* (in press).

Whitley, Darrell, Tim Starkweather and D'Ann Fuquay (1989). Scheduling problems and traveling salesman: the genetic edge recombination operator. *Proceedings of the Third International Conference on Genetic Algorithms,* Palo Alto, Ca.: Morgan Kaufmann. pp. 133-140.

Part III

GENESIS and OOGA: Genetic Algorithm Software

23

Concerning GENESIS and OOGA

Lawrence Davis and John J. Grefenstette

The third part of this book consists of two genetic algorithm software systems and their documentation, available to you by mail on a single diskette.

We retain rights to the software in these systems, although you are free to modify them for your own use. We ask that you not copy the software for multiple use. The modest price we are charging for this software will support further development of these systems—a process we believe will ultimately be of benefit to the genetic algorithm community. We can provide site licenses to educational institutions; write us if you are interested.

We describe GENESIS and OOGA in more detail in the following sections.

CONCERNING GENESIS

GENESIS was written by John Grefenstette to promote the study of genetic algorithms for function optimization. GENESIS has been under development since 1981, and has been widely distributed to the research community since 1985. The system provides the fundamental procedures for genetic selection, crossover and mutation. The user must provide an evaluation function which returns a value when given a particular point in the search space. GENESIS has served as the basis for many of the genetic algorithm results reported since 1985.

The version of GENESIS on the diskette has been updated especially for this book. The improved user interface should increase the ease with which you can tailor GENESIS to new problems. Source code for GENE-SIS is contained on the diskette. The system is written in the language C, and has been successfully installed on a wide variety of machines. Details concerning installation and the interface between the user-written function and GENESIS are explained in the documentation included with the software. UNIX (TM) shell files are provided to ease the construction of genetic algorithms for the user's application.

GENESIS was designed to encourage experiments with genetic algorithms. The user, with a little experience with C programming, should find it easy to investigate new variations of genetic algorithms (e.g., new crossover or mutation operators) by making minor changes to GENESIS.

CONCERNING OOGA

OOGA is a simplified version of the Lisp-based software used by Lawrence Davis over the past seven years. OOGA was created especially for this book. It was designed so that the points made in the tutorial would be clear in the code. OOGA is a system that is easy to tailor to one's own problems and so it may be useful to Lisp users who wish to understand the tutorial more fully, experiment with some workbench problems, or apply the software to more realistic domains.

OOGA runs in Common Lisp and CLOS[1]. It will not run in Common Lisp without object-oriented support and it will not run in the Flavors system[2]. PCL is a portable implementation of CLOS available for all major Common Lisp implementations (usually from the Lisp vendor). By the end of 1990, most Lisps will come with a highly-efficient native CLOS implementation. Dan Cerys has tested the tutorial software on the following machines: Symbolics Lisp Machines, Texas Instruments Lisp Machines,

---

[1]CLOS = Common Lisp Object System: a powerful object-oriented extension of Common Lisp that will be part of the ANSI Common Lisp standard

[2]an older object system that was one of the motivations for CLOS

Apple Macintoshes (with Macintosh Allegro Common Lisp), and a variety of UNIX machines with Franz Allegro and Lucid Common Lisp. If your system is similar to one of these, we believe that you will not have difficulty with OOGA.

HOW TO ORDER GENESIS AND OOGA

To order GENESIS and OOGA on a diskette, together with documentation on their use, fill out the form on the next page and send it with payment to:

TSP - The Software Partnership
P.O. Box 991
Melrose, MA 02176
USA

When you send your check please fill out and enclose the form below telling us the type of diskette you prefer, the computer environment you intend to use, and which of the two systems you plan to use. We will forward relevant information about system enhancements to you on the basis of this data unless you tell us not to do so. Finally, our legal folks want you to know that the software comes with an agreement that releases us from any claims of liability caused by failure of the system, inadequate documentation, and so on.

GA Software Order Form

Please send me _____ copies of the GENESIS and OOGA software, documentation, and single-user license. Price is $52.50 per copy; $60 per copy for addresses outside of North America.

Format of Diskette:
☐ $3\frac{1}{2}$ Macintosh   ☐ $3\frac{1}{2}$ IBM PC   ☐ $5\frac{1}{4}$ IBM PC

Form of Payment: (Checks are preferred)
☐ Check or Money Order (U.S. funds payable to TSP)
☐ Visa   ☐ MasterCard
Card Number: _____
Expiration Date: _____
Name On Card: _____
Signature: _____

Which system(s) are you interested in? (You will receive both systems, regardless of your response).
☐ GENESIS   ☐ OOGA
Type of computer you intend to use: _____ .

Why are you interested in this software?
☐ Education (self-study)   ☐ Education (classroom)
☐ Applications   ☐ Research

Shipping Address:
Name: _____
Address: _____
        _____
        _____
Daytime Phone: _____
EMail Address: _____

Send to: TSP (The Software Partnership)
        P.O. Box 991
        Melrose, MA 02176
        USA

Contributing
Authors

M. J. J. Blommers
Laboratory for Analytical Chemistry
Faculty of Science
Catholic University of Nijmegen
Toernooiveld 1
6525 Ed Nijmegen
The Netherlands

Eugene E. Bouchard
Lockheed Aeronautical Systems Co.
PO Box 551
72-53 Plant 2 Unit 50
Burbank, CA 91520

Mark F. Bramlette
Lockheed Aeronautical Systems Co.
PO Box 551
72-53 Plant 2 Unit 50
Burbank, CA 91520

L. M. C. Buydens
Laboratory for Analytical Chemistry
Faculty of Science
Catholic University of Nijmegen
Toernooiveld 1
6525 Ed Nijmegen
The Netherlands

Louis Anthony Cox, Jr.
U S West Advanced Technologies
6200 S. Quebec St.
Englewood, CO 80111

Yuval Davidor
Faculty of Mathematical Sciences
The Weizmann Institute of Science
Rehovot 76100
Israel

378

Lawrence (David) Davis, during the preparation of this book
Bolt Beranek and Newman Systems and Technologies Corp.
10 Moulton St.
Cambridge, MA 02138

Lawrence (David) Davis, at present
Tica Associates
36 Hampshire St.
Cambridge, MA 02139
(617) 864-2292

Stephanie Forrest, during the preparation of this book
Center for Nonlinear Studies
MS-B258
Los Alamos National Laboratory
Los Alamos, NM 87545

Stephanie Forrest, at present
Dept. of Computer Science
University of New Mexico
Albuquerque, N.M. 87131

John J. Grefenstette
Naval Research Laboratory
Washington, DC 20375-5000

Steven A. Harp
Honeywell CSDD
1000 Boone Ave. N.
Golden Valley, MN 55427

Charles Karr
U.S. Bureau of Mines
Tuscaloosa Research Center
U. of Alabama Campus
PO Box L
Tuscaloosa, AL 35486-9777

G. Kateman
Laboratory for Analytical Chemistry
Faculty of Science
Catholic University of Nijmegen
Toernooiveld 1
6525 Ed Nijmegen
The Netherlands

Gunar E. Liepins
Oak Ridge National Laboratory
Oak Ridge, TN 37831

C. B. Lucasius
Laboratory for Analytical Chemistry
Faculty of Science
Catholic University of Nijmegen
Toernooiveld 1
6525 Ed Nijmegen
The Netherlands

Gottfried Mayer-Kress
Center for Nonlinear Studies
MS-B258
Los Alamos National Laboratory
Los Alamos, NM 87545
and
Department of Mathematics
University of California at Santa
Cruz
Santa Cruz, CA
and
Santa Fe Institute
Santa Fe, N.M.

David Montana
BBN Systems and Technologies
Corp.
70 Fawcett St.
Cambridge, MA 02138

W. D. Potter
Department of Computer Science
University of Georgia

David J. Powell
Corporate Research and Development
General Electric Company
Schenectady, NY

Yuping Qiu
U S West Advanced Technologies
6200 S. Quebec St.
Englewood, CO 80111

T. Samad
Honeywell CSDD
1000 Boone Ave. N.
Golden Valley, MN 55427

Daniel Shaner
Hewlett/Packard
Fort Collins, CO 80525

Michael M. Skolnick
Department of Computer Science
Rensselaer Polytechnic Institute
Troy, NY 12180

Timothy Starkweather
Computer Science Department
Colorado State University
Fort Collins, CO 80523

Gilbert Syswerda
BBN Systems and Technologies
Corp.
10 Moulton St.
Cambridge, MA 02138

Siu Shing Tong
Corporate Research and Development
General Electric Company
Schenectady, NY 12301

Darrell Whitley
Computer Science Department
Colorado State University
Fort Collins, CO 80523

Index

0–1 integer programming problems, 238, 240, 244
2 1/2 D sketch, 284

adaptation, 282, 285, 287, 294
adaptation of operator fitness, 95
adapting operators, 57
adaptive parameterization, 92
addition operator, 155
aircraft engine turbine design, 327
analogous crossover, 152
arm-configuration, 147
artificial intelligence, 238, 243
average crossover, 66

backpropagation, 301, 306
biased mutate weights operator, 304
binary f6, 8
binary representation, 11

bit mutation rate, 12
bit string encoding, 62
blackboard system, 284
building block, 19

CAP, 238–240, 243, 244, 246–248
cascaded hybrid evaluation, 266
causal, 247
causal relations, 238, 316
causality, 245
children, 12
chromosome interpreter, 338
chromosome, lists, 337
chromosomes, 2
classifier systems, ii, 239
clustered crossover, 195
combinatorial optimization, 72, 124, 336
common cause faults, 238
communication, 238, 248

Communication Alarm Processor Expert System, 237
conformation, 256
conformational analysis, 251, 252
conformational space, 256
connectionist, 238, 248
constrained optimization problems, 241, 242, 245, 248
constraints, 333
control problem with time pressure, 126
convergence, 25, 118
credit assignment to operators, 95
crossover, 16, 291, 303, 342
crossover features, 305
crossover nodes, 305
crossover probability, 245
crossover rate, 12
crossover weights, 305
current algorithm, 56
current encoding, 56

decision trees, 239, 240
decoding, 57
DeJong test suite, 91
delete last, 36
deletion operator, 155
deletion techniques, 118
DENISE, 251
direct kinematics, 147
distributed evaluations, 287, 296, 298
distributed genetic algorithm, 355
DNA hairpins, 258
DNA structure, 253
domain-dependent rules, 329
domain-dependent search, 318
domain-independent search, 321
dynamic routing policy, 128

edge recombination, 352
edge recombination crossover, 343
edge recombination operator, 351
elitist, 245
encoding techniques, 4
end-effector, 147
epistasis, 157
evaluation function, 4
evaluation module, 8
evaluations, number of, 347
expert systems, 282, 283, 314, 318
exploitation, 315

F-14, 333
feasibility, 242
feasible, 242
figure-ground separation, 284
fitness function, 173
fitness is evaluation, 11
fitness techniques, 32

GA-hardness, 179, 268
generalized reduced gradient method, 315
generational replacement, 11, 40
GENESIS, vi, 175
GENITOR, 35, 352
graded penalty, 245
graph coloring problem, 72, 73
Gray coding, 92, 116, 174, 264
greedy algorithm, 73

HASP/SIAP, 284
heuristics, adapting, 58
hill-climbing, 10, 114, 181
hillclimb operator, 305
Holland, 2, 262
homologous cross site, 151
hybrid genetic algorithm, 58, 72

hybrid optimization techniques, 138
hybridization, 54, 57
hybridization principles, 56
hybridization techniques, 77
hybridizing genetic algorithms, 61
hyperplane sampling, 369

induction, 239
initialization techniques, 118, 315
initializing operator fitness, 96
interdigitation, 313, 316
interdigitation of expert system and numerical optimization, 318
inversion, 21

knowledge engineering, 238, 239, 248

Lamarckian probability, 161, 163
Lamarckian probability crossover, 158
Lamarckism, 156
learning, 239, 240, 248, 282, 285, 295

mass mutation, 246
mathematical programming, 238
microwave communication diagnosis, 238
microwave communication faults, 237
migration, 355
modified method of feasible directions, 315, 326
multifault diagnostic problems, 239
multiple-fault diagnosis, 237, 238, 243, 248

mutate nodes operator, 304
mutate weakest nodes operator, 304
mutation, 2, 245, 291, 303, 340
mutation probability, 245

natural selection, 2
network control problem, 124
neural network training, 282, 301
neural networks, 239, 240, 263, 282, 301
NOE spectrometry, 259
nonlinear constrained optimization, 315
nonlinear dynamical systems, 166
nonmonotonic reasoning, 238
Norms, 170
NP-completeness, 77, 125, 238
numerical optimization, 313, 314, 321
numerical optimization and genetic algorithms, 321
numerical representation, 62

one-point crossover, 16
one-point crossover and mutate, 12, 15
OOGA, vi, 28, 289
operator performance computation, 93
operator probability adaptation, 292
operator weights, 12
operator, syntactic, 336
operator-based reproduction, 45
operators, 339
operators, mix of, 345
opportunistic pairwise exchange, 136
optimization techniques, 314
order-based chromosomes, 134

order-based crossover, 343
order-based mutation, 340
order-based representation, 77

parallelism, 287, 296, 298, 299
parameter tuning, 282, 285, 310
parameterization of genetic algorithms, 38, 91, 120
parameterization of nonlinear dynamical systems, 167
parametric design, 110
parents, 12
Pareto optimal, 289
path-following programs, 147
penalty functions, 242, 243, 248
permutation encoding, 78, 117, 337
permuted list representation, 78
planning, 332
population size, 347
position-based crossover, 343
position-based mutation, 340
prisoner's dilemma, 170
probability of detection, 288, 290
probability of false alarm, 288, 290, 294

random binary initialization, 11
random generate and test, 61
random initialization, 118
random search, 114, 137, 340
real number creep, 66
real number mutation, 66
recombination, 2, 242, 245
recursive pairwise refinement, 270
reproduction, 2
Richardson model, the, 168
robot trajectories, 148, 150, 159, 163
robustness, 39, 40, 55, 70, 92

ROC curves, 288, 289, 291, 294
roulette wheel parent selection, 11, 13
rule mutation operator, 196
rule threshold optimization, 288, 296
rules, 282, 288, 294

scaling fitness techniques, 194
schedule editor, 334
schedule optimizer, 334
scheduling, 125, 332, 350
schema theory, 19, 353
scramble mutation, 340
seeding, 315, 322
segregation crossover, 153
sensitivity analysis, 246, 247
sequence, 256
sequential linear programming, 315
sequential quadratic programming, 315, 326
set-covering, 240–245, 248
sharing, 37, 268
signal detection, 282, 284, 295
simulated annealing, 114, 262
single-fault diagnosis, 237
specialize operator, 193
steady-state genetic algorithms, 292
steady-state reproduction, 35
steady-state without duplicates, 36
stochastic hill climbing, 117
structure, 256

TABU search, 243
telecommunications network, 125
test set of engineering problems, 324

three-agent Richardson's model,
        168
tracking, 282, 295
traveling salesman problem, 351
traveling salesperson problem,
        337
two-opt, 136
two-point crossover, 48, 245

unbiased mutate weights opera-
        tor, 304
uniform crossover, 49, 291

visual texture, 284

Watson–Crick DNA, 258
windowing, 290, 292